Intuitive Math:

100+ Power Strategies for ACT® and SAT® Math

Advanced Skills for the Math Anxious and Math Gifted

JAY B CUTTS, MA

Director, Cutts Graduate Reviews
Albuquerque, NM

ACT® and SAT® are trademarks registered by ACT, Inc. and the College Board, respectively, which are not affiliated with, and do not endorse, this product.

ABOUT THE AUTHOR

Jay Cutts is the director of the Cutts Graduate Reviews. For over 30 years he has helped students master the math and verbal skills needed to perform well on entrance exams for college, graduate school, law school, medical school, business school, and pharmacy school. He specializes in advanced, personalized, and intuitive strategies for test taking, timing, math and verbal problem solving, and math and test anxiety.

He is the lead author of the Barron's® LSAT® and MCAT® prep books and the Barron's® MCAT flash cards. He has also published a number of works of fiction, including *Death by Haggis* and *Annie Gomez and the Gigantic Foot of Doom*.

In his spare time he gardens, dances, and plays the accordion.

© 2020 by Jay B Cutts

All rights reserved.
No part of this publication may be reproduced or distributed in any form or by any means without the written permission of the copyright owner.

All inquiries should be addressed to:
Jay Cutts
31 B Kelly Lane
Los Lunas, NM 87031
www.cuttsreviews.com

ISBN: 978-1-7346306-0-2

ACT® and SAT® are trademarks registered by ACT, Inc. and the College Board, respectively, which are not affiliated with, and do not endorse, this product.

*Barron's® is a trademark owned by Barron's Educational Series, Inc., which is not affiliated with, and does not endorse, this product.

®LSAT and ®MCAT are registered trademarks of Law School Admissions Council, Inc. and the American Association of Medical Colleges, respectively, which do not endorse this product.

Edition: 1

Version 1.0

Contents

Preface ………………………………………………………………………………….v

Chapter 1. Introduction…………………………………………………..7

What is Intuitive Math?
The Problem-Solving Process
How to Use This Book

Chapter 2. General Strategies……………………………………………11

Timing Strategy
Calculator
Setup Strategies
Organizing Strategies
Problem-Solving Strategies
Calculation Strategies

Chapter 3. Specific Patterns and Strategies…………………………….17

Calculations
Averages
Circles and Triangles
Ratio, Rate, and Percent
Exponents, Roots, and Logs
Equations and Inequalities
Quadratic Equations
(x,y) Coordinates

Chapter 4. ACT®-style Questions with Explanations……………….41

Chapter 5. SAT®-style Non-calculator Questions with Explanations……..143

Chapter 6. SAT®-style Calculator Questions with Explanations………….175

Index of Problem Types………………………………………………213

Bonus Material……………………………………………………..223

Excerpt from *Annie Gomez and the Gigantic Foot of Doom*

Preface

This book is the product of over thirty years of working with students on math problem-solving for standardized exams.

I have had students who had superior math skills and wanted to improve their natural intuitions. I have had students who could barely look at a math problem without experiencing physical and emotional discomfort or even paralysis.

But every student with whom I have worked has learned that there is a vast, and perhaps unlimited, reservoir of intuitive strategies for math into which they could learn to tap. And every student has come away from the process with increased confidence in their ability to understand, organize, and solve math problems.

Math is about relationships. Physical, spacial, temporal relationships. Relationships in two dimensions and in three dimensions. Abstract relationships and very hands-on, practical relationships. Every student with whom I have worked has had a sense of these relationships. Seeing math as relationships is especially important to those students, young or older, who are people-oriented and fear that math is too "intellectual" for them.

Standardized exams pose particular problems to all students, whether they are strong in math or struggle with it. These exams force students to work faster than they are comfortable with. They emphasize problem solving rather than the memorization of math processes that many students have to study for their math classes.

It is my hope that, by sharing what I have learned about intuitive math over 30 years, I can help any student who needs to take the standardized entrance exams to learn tools they can actually understand and use to increase their score and decrease their anxiety.

As one adult student once exclaimed to me, "I can do this! I can understand it! How come no one ever taught me math like this before?"

This book is dedicated to each and every one of you who has ever puzzled over math, either with keen interest or with anxiety. Life is a vast, intuitive, mysterious process. Math is one doorway that you can learn to use to go forth into new horizons! May you learn, explore, and grow.

If you have questions about the material in this book, would like to report possible math or typing errors, or have suggestions for making the book more effective for students like yourself, I would be glad to hear from you. You can contact me at orders@cuttsreviews.com.

At the end of the book I have included an excerpt from my novel ***Annie Gomez and the Gigantic Foot of Doom***, a story about how ordinary and unordinary high schoolers save humanity (again). Hopefully, that will serve as a reward for finishing the book!

Jay B. Cutts, MA

Acknowledgements

I would like to thank all of the students with whom I have worked over the past 30 years for helping me understand what it takes to convey intuitive math strategies to students who are struggling and to students who are eager to expand their math skills.

I want to thank all of the editors with whom I've worked in my publishing career. I especially want to acknowledge Linda Turner, formerly with Barron's Educational Services, for her patient coaching and her superior editing and production skills, as well as for being a friend.

I want to thank whomever it was that left a dog-eared algebra book in my closet when I was in sixth grade. Bored with sixth grade math, I picked up that fascinating book and began an exploration into the complex world of math relationships.

Thanks to all of my test prep friends and colleagues for their support and companionship over the past 30 years. A special acknowledgement to John F. Mares, for introducing me to test preparation, to Janelle Campbell and our late friend Jack Oberst for introducing me to John Mares and to the world of test prep, and to Scott Hildebrandt of eKnowlege, for encouraging me to make my years of intuitive math insights accessible to ACT® and SAT® students.

Thanks to my friends and my family. As years pass, it is increasingly clear that they are what life is about.

To Baxter, Ruby, and Kodi, who are furry and affectionate, thanks for the wags and purrs.

Finally, my deepest appreciation to my life partner Linda for her unwavering support, not only as I wrote this book but in all things.

Chapter 1. Introduction

- What is Intuitive Math
- The Problem-Solving Process
- How to Use This Book

What is Intuitive Math

The purpose of this book is to help you learn what we call "intuitive" math strategies. Intuitive strategies draw on tools that anyone can learn, understand, and use. Intuitive strategies use pictures, charts, road maps, and other hands-on tools that help you organize math information in a way that feels concrete and understandable to you.

Many math classes teach math in an abstract way. You memorize certain processes for solving certain kinds of problems. You may not have a chance to really understand why or how those processes work. Algebra, trigonometry, (x,y) coordinates, squares and square roots, quadratic equations can all become so abstract that you can easily get lost and confused when you work on them.

All of math is based on relationships. Intuitive math gives you tools for understand these relationships. Once you can understand a relationship, you can start solving problems in a way that makes sense to you.

I was once working on a math problem for fun (yes, it can happen!) and was just about to put my pencil to the paper and start doing some calculations when I realized something revolutionary. I realized that I had just spent the previous five minutes orienting myself to the relationships, organizing the information in my mind, and considering various possibilities for attacking the problem. All of this was a completely intuitive process. There were no rules or formulas that I was following. And I had hardly even been aware that I was doing it!

For people who are naturally good at math, there is a whole mysterious world of math intuitions that they naturally plug into. They may not be aware they are doing it. They take it for granted. If someone asks them to explain a math question, they will fast forward to the standard, abstract problem-solving steps without sharing their instinctive intuitions with the other person, because they are not aware of their intuitions. But the intuitions are the **only** thing that allows them to solve the problem. Without the intuitions, all of the formulas and techniques in the world will not help solve a math problem.

Math intuitions are learnable. Anyone can learn them. And understand them. And apply them. As you go through this book, attempting problems and reading the explanations, everything you will be exposed to is an intuitive strategy.

Keep your eyes open. Try out new strategies. If you feel frustrated, take a breath, take a break, and then come back to the problem. The key to intuitive math strategy is that it has to make sense to **you**. If a certain strategy does not work for you, look for your own strategies. You will find them!

The Problem-Solving Process

The standardized entrance exams that you have to take test math problem solving more than memorization of math facts. What is problem solving?

In this book we will be looking at hundreds of good problem-solving tools. To understand problem solving, it is helpful to look at it in terms of your thinking styles. People have two distinct thinking styles. We can call them wholistic versus detail-oriented. We could also call them Big Picture thinking versus detail thinking.

These correspond to the very different ways of functioning of the left side of your brain and the right side of your brain. The right side (which controls the left side of the body) processes information in a "Big Picture" way. It sees the whole story. It is intuitive. It understands relationships. It gets the overall idea.

The left side of the brain (which controls the right side of the body) focuses on details, specifics. It is very good at that but easily loses sight of the Big Picture. It applies rules but does not necessarily remember why. It easily gets lost in detail.

Problem solving requires coordinating both processes. Neurologically, we probably cannot use both Big Picture thinking and detail thinking at the same time. We have to learn to go back and forth between them. As you work on the problems in this book, you can start to notice when you are getting lost in detail. If you notice it, you can back off a little bit and draw on the more wholistic strategies that you are learning.

It is also possible that you might get lost in wholistic processing. You see the problem. You get the picture. But you have no idea of where to start getting to the answer. Many intuitive strategies help you focus on the details.

When you are working on a problem and you feel stuck, with no idea of what to do next, you have probably come to the end of either Big Picture thinking of detail thinking. Take a breath and remember the tools you are learning here to switch to a different perspective.

How to Use This Book

I have tried to make this book as hands-on as possible. You can probably learn best by going through problems and reading through the explanations. However, the next two chapters will present and review some important strategies for you in a more "lesson-like" way. Read through those chapters and get what you can out of it. However, if something does not make sense or is frustrating, you might find it easier to learn as you go through the problems in the book.

Most of the patterns that you will see on your standardized test are covered in the 101 math questions here. If you work through the problems and explanations carefully and patiently, you will learn the tools you need without having to work too hard at it. Learning hands-on is easier, more effective, and more pleasant than trying to study lessons!

Start out by reading through Chapters 1 through 3. Chapter 2 does not contain math strategies but rather focuses on important general test-taking strategies. Do read these carefully. Chapter 3 is a quick review of the major types of problem-solving tools. You do not need to get too bogged down in these. You will learn all of them as you go through the problems. However, it might help you to read through the information in Chapter 3 and let it sink in a bit.

The rest of the book consists of ACT®-style questions, SAT® Non-Calculator-section-style questions, and SAT© Calculator-section-style questions. No matter which test you take, **all** of these questions will help you build intuitive strategies. The formats are slightly different but the content is mostly similar. If you only study the questions for the test you are going to take, you will be missing out on a lot of important patterns and strategies.

The sample questions are divided into 20 daily assignments, each covering about 5 questions. You can do one day's assignment at a time or you can go faster or slower. After each daily assignment, you can keep track of how you did and what you learned.

For each question, you will probably get the most out of it if you try to solve the question on your own before reading the explanation. Give yourself as much time as you need. If you want to spend half an hour or an hour on one question, great! The more you wrestle with a question, the more your own intuitions will begin to work.

After you have tried your best, read through the explanation, even if you got the right answer! This is very important because getting the right answer is not your primary goal. Your goal is to understand what you are doing and to develop the most effective intuitive tools. The explanations are designed to help you do that.

In writing explanations, I have aimed at being very thorough, so that everyone, regardless of their math background, can break the problem down. If something in an explanation is already very clear to you, you can skip it. Focus on the strategies that are new. However, it might be a good idea to read through even the explanations that you feel you already are clear on. Even simple tasks can be done in more effective ways.

If you are very strong in math, you may already be able to get many of the questions correct. The challenge for you on the actual exam is accuracy and timing. You may have tools for getting to the right answer but if your tools are more abstract, you can easily make a number of careless errors. Your goal is to learn new intuitive tools that will help you double-check your work and maintain your accuracy. Intuitive tools will also save you time because they are more direct and less prone to error and confusion.

The intuitive insights in this book will also help you get certain more complex questions right that you would not get right otherwise. The standardized exams test problem solving more than math memorization. Focus on learning intuitive tools for becoming a more creative problem solver.

If you wrestle with math, you will find a lot here to learn. Maybe it will get overwhelming at times. But you can always take a break and come back later. Most of all, you will find that intuitive tools are tools that you can understand and use. You will start solving problems that looked impossible before.

If something in an explanation does not work for you, that is okay. I have tried to keep the explanations organized enough that everyone can follow them, but sometimes an explanation may just get overwhelming. I apologize for that. Try coming back to it later. Try finding your own intuitive way to understand and solve the problem. Try discussing the problem with a friend.

At the end of the book is an index of major problem types, showing you in which problems that type appears. This index can help you find problems for practicing certain kinds of math. It also helps you get an overview of the main types of problems that your standardized exam will test.

About the Drawings and Diagrams. For legibility, most of the diagrams and drawings in the book have been made electronically. When you use these visual aids in your own practicing and testing, you will of course have to write them out by hand yourself, so they won't look exactly like the diagrams here. This includes making tables, drawing diagrams, and writing out road maps. When you practice, use these visual aids as much as possible. Get used to finding your own way to draw them.

Connect through Twitter. To share with others who are working on intuitive math, join our Twitter list at:

https://twitter.com/i/lists/1267567938288009217

Specific Questions or Comments? If you have a personal question or comment for Jay Cutts, please email orders@cuttsreviews.com. You are welcome to let us know about any errors in the book or any suggestions you have for making the book more helpful.

Chapter 2. General Strategies

- Timing Strategy
- Calculator
- Setup Strategies
- Organizing Strategies
- Problem-Solving Strategies
- Calculation Strategies

This chapter covers important general strategies. Do read this carefully before going on to the next chapter or the test questions.

Timing Strategy

How you use your time on the test is really critical. Your standardized exam is very different from tests you take in class. You have been trained to take tests a certain way but you will lose a lot of points on your standardized exam if you try to apply the same approach.

One problem is that the timing strategy that you have developed is unconscious. You probably cannot describe how you make decisions about using your time on tests in class. You just do it automatically. For that reason, even if you study a different way to use your time for the standardized exams, your automatic training will probably kick in and sabotage you.

To overcome your automatic training it can be helpful to work with another person on timing, especially if you have access to someone who is familiar with the test. Below are some basic guidelines for you to start experimenting with.

Your automatic approach to test taking most likely involves spending about a minute on a question and then, if you do not have the answer, guessing and moving on. You probably aim at answering all the questions on the test. You probably start with the first question and then go in order.

On a test in class, if you do not get at least 85% of the questions right, you probably will be unhappy with your score. On the standardized exams you can probably get a competitive score even if you get just half of the questions right. Obviously, if you can get more right, you should, but you should not panic if you are only getting part way through the exam. If you rush to get to more questions, you will get many of them wrong.

It is better to work on a little bit fewer questions – putting cold guesses down for the ones you do not get to – and get a higher percentage of the questions that you work on right.

Consider a person who works on all 20 questions in a certain math section. Their results might be 6 questions right and 14 questions wrong. Suppose that person had cut 5 of the 14 wrong questions altogether, just putting cold guesses down for them. That would have given them an extra 5 to 8 minutes that they could have used on the remaining 9 of the wrong questions. That would be enough extra time to get maybe another 5 questions right.

Instead of working on 20 for a score of 6, they would work on 15 for a score of 11. Cutting some questions so that you can spend a little more time on more difficult questions increases your total score.

How much time is ok to spend on a question? Your test taking experience has probably left you feeling anxious if you do not have the answer in 60 to 90 seconds. On regular tests, if you do not know the answer right away, you probably will never get it. However, on the standardized entrance exams you are being tested on problem-solving ability. If you do not see the solution right away, you have tools you can use to keep working.

It is probably ok to spend up to 3 minutes on a question if you need to. Even 4 minutes might be ok if you are getting close to an answer. It is probably **not** a good idea to spend 6 minutes or more. You can experiment with this for yourself.

How can you tell if your timing strategy is good? If most of the questions you work on are right, you are using your time well. If you have let yourself spend 3 minutes on the questions you get wrong, your timing is good. On the other hand if you are getting half the questions you work on wrong or if you are only spending a minute on the questions you get wrong, you are not using your time as well as you could.

In order to use your time well, you have to be able to keep track of time during your test. The tests generally have a clock on the computer. When you begin a question, you must write down the start time. For most questions you will get to an answer in about a minute and you can then move on. However, if you are starting to feel that you have spent too much time on a question, you have to be able to check the time and figure out quickly how much time you have spent. If it is only a minute or two, you should spend more time.

If you do not keep track of time when you start a question, you will not know how long you have spent on it. If you start to feel like the question is taking a long time, you will probably just guess on it and get it wrong because you did not know how much time you had spent.

Keeping track of time and learning to allow yourself to spend a little more time takes a lot of practice. You have to evaluate your results honestly. However, it is very much worth it. Many of my students have reached their goals **just** by learning how to use their time more effectively.

Another aspect of timing is to choose what you work on. Very few students can effectively work on all the problems in a section. Almost everyone will do better by cutting some questions. That means you can choose what to work on. Skip the questions that will clearly be challenging. For example, if you hate sines and cosines, skip those questions.

If you simply start at the beginning and work until time is up, you will be working on both easy and hard questions. If you make an effort to choose what to work on, you will work on easier questions and automatically increase your score.

Calculator

The ACT© currently allows you to use a calculator on the math section. The SAT© currently has one section on which you cannot use a calculator and a second section on which you can.

Should you use the calculator where it is allowed? It depends on your skill and confidence in your results. If you are highly accurate with the calculator, it can save you time. If you tend to make mistakes, you should, at the very least, use intuitive calculation tools for double-checking your work. Or you should skip the calculator altogether and learn intuitive tools for doing calculations with a high degree of accuracy.

Intuitive calculation tools sometimes seem simplistic but they are very powerful because of their accuracy. This book presents many tools for doing various kinds of calculations. You may or may not decide to use them but you should definitely study them. A high percentage of wrong answers are due to incorrect calculations. In other words, the test taker knew how to do the problem but got no credit because their calculations had an error.

Setup Strategies

Your first task in any question is to set up the information so that you can understand it, organize it, and problem-solve with it. Most people tend to simply read the question and jump right into the answer choices. It is helpful to break this process down into its important parts.

Orient to the question stem. The question stem is the part of the question before you get to the answer choices. Instead of just reading the stem and going on to the answers, it is very important to take some time for orienting yourself to the problem. Many wrong answers are due to the fact that the test taker never really understood the question.

Orienting means getting a Big Picture sense of what the problem is about. Ask yourself what the components of the problem are. What are the relationships? What are you being asked to do? There is no point continuing with the problem if you are unclear as to what you are dealing with.

Later on, as you are working on the problem, come back once in a while to the question stem and reread it. Often, people forget what it was they were supposed to answer and end up getting the question wrong.

Look for what will be confusing. Part of orienting is spotting the elements that the test makers hope will confuse you. Negative numbers are more confusing than positive ones. Fractions are more confusing than whole numbers. Subtraction and division are more confusing than addition and multiplication. By spotting elements that can be confusing, you are better prepared to be careful and avoid falling into traps.

Multiple steps make a problem complicated. If a problem has multiple steps to it, it will be easy to get lost or confused. Spotting this up front helps you be prepared to organize the steps systematically.

Review the answer choices. At this stage it can be helpful to just glance at the answer choices. This gives you an idea of what you have to distinguish and what you do not. For example, if answer choices are 3, 30, 300, 3000, then you do not have to worry about decimal points or small differences. You just need to know a general ballpark figure. Noticing the similarities and differences between answer choices helps you focus in on what you need to figure out.

Generally, numerical answer choices are in an ascending or descending order. If the answer choices are getting bigger and choice C is already too big, then choices D and E are also too big. Some problems, however, purposely do not have answer choices in order

Must be true. Some question stems ask for an answer that "must be true." This wording can be confusing. If the correct answer must be true, what does that mean about the wrong answers? That they must be false? No! The wrong answers are not necessarily answers that cannot be true. They may be answers that could be true but do not have to be.

Organizing Strategies

After setting up a problem, your next task is to organize the information. There are far too many organizing tools to list here – you will learn them as you go through the problems – but there are two important tools that should be mentioned.

Road maps. A road map is a plan of attack that you actually write out on your scratch paper. Creating a plan of attack is a Big Picture process. Once you dive into the details of a problem, it is very likely that you will lose track of the Big Picture and forget what you are doing.

To avoid this, take the time as you set up the problem to think through your plan of attack and write it down in steps. It may seem overly simplistic when you do it but it will save you later on when you have gotten bogged down in detail thinking.

You do not need to write down a road map for every question. Only the more complex questions need a road map. You have to find your own balance for when to write one and when not to. If you are getting questions wrong because you have gotten confused about the steps you need to take, use road maps more often.

In addition, the act of writing the road map down can help you organize your thoughts.

Use tables and visual aids. This is not so much a tool as a Commandment! Visual aids are the key to organizing most problems. If you try to organize information in your head, your processing power is quickly overwhelmed and you make huge mistakes. Even if your brain is still fresh, trying to process information in your mind is not accurate. The minute you put information into your mind, it starts mutating. Was that ten people or eight people? Wait, who went first, Robert or Sandy?

Your most powerful tool is to get things on paper. In this book you will learn dozens of ways to do this for various types of problems.

Problem-Solving Strategies

Once you have oriented yourself to the problem and organized the information, you are ready to apply your problem-solving skills. You will learn these skills by going through all the problems in this book. Below are a few specific tips to remember.

The KING approach. This is one of the most powerful problem-solving strategies. In this approach you ask yourself a series of questions that help your brain engage in problem solving. KING stands for

K – What do I Know
N – What do I Need to now to solve the problem
I – What can be Inferred from what I know
G – How do I Get the information I need.

By consciously identifying what you already know, you prep your brain for organizing the information. By asking yourself what you need to know, you identify what to focus on. By asking what can be inferred from what you know, you apply logic to gather more information. By asking how you get the information you need, you identify a path to the solution.

The KING approach is especially helpful when you are stuck on a problem and don't know what to do. By asking yourself the KING questions, you are likely to get new insights into how to solve the problem.

Test answer choices. Generally, there are two ways to approach solving a problem. The most common is to do some math manipulations to figure out the answer and then find the answer choice that matches it. Sometimes, though, it is easier just to test which answer choice might work. In many cases you can get part way to an answer, or eliminate certain answer choices, by math manipulations and then can get the rest of the way to the correct answer by testing out the remaining answer choices.

As an example, suppose a question asked you to solve for x, given $3x - 12 = 3$. The first method is to manipulate the equation until you have solved for x. However, if the answer choices included 3 and 5, you could just test them without having to change the equation. 3*3 – 12 does not equal 3. But 3*5 –12 does. You are done!

Assign a value to a variable. In some problems that have variables in the answer choices, you can assign a value to the variable to help you solve the problem.

1. Which of the following expressions is equivalent to x^2

A. $4x$

B. $x * |x|$

C. $|x| * |x|$

D. $-x * x$

If you assign the value 3 to x, you can then simply plug 3 into the original expression, getting 9. Then plug 3 into each answer choice. Any choice that does not equal 9 is out. However, in this method, more than one choice might work. If that happens, any answer that was out stays out but you have to assign a different value of x to test the remaining choices.

In this case choice A is 16 and is out. Choice B is 9. Choice C is 9. Choice D is –9 and is out.

If you now assign –3 to x, choice B is –9 and is out. Choice C is 9 and is the correct answer.

Process of elimination. In the example above you were able to eliminate two answer choices. In the second phase you were able to eliminate another, thus finding the answer by process of elimination. This strategy helps you focus on what is left. It can be easy to forget to use process of elimination. Train yourself to remember it.

Check to see if you already have enough information. People usually try to solve a problem all the way to a complete solution. Often this is not necessary. Train yourself to check from time to time to see if you already have enough information to answer the question, or at least to eliminate some of the answer choices.

Calculation Strategies

There are many powerful strategies for carrying out calculations more intuitively and thus more accurately. You will learn dozens of these as you go through the problems in the book. Below are two basic principles for intuitive calculations.

Regroup numbers. If you have a complicated string of additions or multiplications, rearranging the numbers into a more intuitive order is a powerful tool. Consider the addition below.

$$15 + 7 + 9 + 3 + 1 + 5$$

$$(15 + 5) + (9 + 1) + (7 + 3)$$

By regrouping, you can do all of the additions in your head accurately – 20 + 10 + 10.

Break calculations into steps. This is a basic and critical strategy. Most of us tend to do calculations in our head. For high accuracy, break calculations down into each tiny step. This may feel tedious at first but when you consider that a high percent of people's mistakes are due to calculation errors, this is a painless way to powerfully boost your score.

The explanations to the questions in this book typically break calculations down into small steps.

Chapter 3. Specific Patterns and Strategies

- Calculations
- Averages
- Circles and Triangles
- Ratio, Rate, and Percent
- Exponents, Roots, and Logs
- Equations and Inequalities
- Quadratic Equations
- (x,y) Coordinates
- Summary

This chapter gives you a brief review of some important intuitive strategies. The best way to learn these is to read through the explanations for the problems in this book. Hands-on learning is more intuitive than reading abstract explanations!

Nevertheless, it might be helpful for you to go through this more abstract review of strategies so that you have a little more background when you read the explanations of problems. If something here seems a little confusing, you might find it easier to understand when you run across the same thing in a problem.

Feel free to play with some of the strategies presented in this chapter. Try them out for yourself. See if you can understand how and why they work. You can even come up with your own strategies.

If you are strong in math, some intuitive strategies might strike you as simplistic. They are not. Intuitive strategies should be simple, not simplistic. To be effective problem-solving tools, strategies must be easy to understand, easy to keep straight, and easy to apply. Yes, you may be perfectly capable of performing complex algebra manipulations, for example. But learning intuitive strategies helps you broaden your base of understanding of how and why the algebra works.

When it comes to a unique math problem-solving situation, a strong mathematician needs to have an extensive repertoire of intuitive tools. In math problem solving, intuition comes first. Math manipulations only come in after there is an intuitive understanding of the problem.

On your standardized exam, if you have good math skills, you can use standard math strategies. However, intuitive strategies allow you to solve math more accurately. Even great mathematicians – and perhaps especially great mathematicians – can set up a problem correctly but often make a careless error.

In addition, intuitive strategies help you solve standardized test questions more quickly. This book points out most of the major patterns around which ACT® and SAT® math questions are built. By studying the strategies for these questions, you will be able to understand a problem more quickly and solve it more quickly. You will know exactly what you are doing and the fastest and most accurate way to do it.

Are the strategies in this chapter the best ones for you? Maybe. Maybe not. An intuitive strategy must make sense to you or it is not intuitive! Try these strategies out. Play with them. Give them the benefit of the doubt. But feel free to come up with other approaches that make sense to you and are clear and simple.

The essence of most intuitive strategies is that they are visual, hands-on, well-organized, and make sense. It is critical to get problem solving out of your head and onto paper or into your hands. Almost everyone tries at first to solve math problems mostly in their head. Check out for yourself why this is a problem. You can notice how easily your memory confuses things, makes mistakes, and just plain forgets what it is doing.

If there is one principle you should get out of this book, it is "put your pencil to the paper." Even if you do not know what to do, even if you are doodling, get your attention onto the paper, where you can organize information without it getting confused, turned around, or forgotten.

> **The Prime Directive**
>
> "Put your pencil to the paper!"

Calculations

Let's look at some intuitive strategies for performing calculations. Many of the wrong answers that people get on the standardized exams are due to calculation errors. It is a shame to know how to do a problem but add a couple numbers wrong.

Be especially careful about all calculations on the test. If you can use a calculator without ever making mistakes, do so where allowed. But if you sometimes get things wrong on the calculator, you should at least double-check your work and possibly even put aside the calculator and use an intuitive process for getting to the right number. The strategies below can help you be more accurate.

Divisible by 3. You may already know the trick for determining if a number is divisible by 3. Add up the digits. If the result is divisible by 3, then the original number is, as well.

$$4{,}235{,}172$$

The digits add up to 24, whose digits add up to 6. The number is divisible by 3.

Change subtractions to additions. It is more confusing to do subtraction than to do addition. If you have to manipulate an algebraic equation that includes subtractions, it is easy to lose track of the subtraction as you move elements around.

The cure is to change subtractions to additions. Consider

$$3x - 17 - 5y = 13$$

To change the subtractions to additions, you can rewrite the left side as

$$3x + \text{-}17 + \text{-}5y = 13$$

Now as you move elements around, the negative sign moves with that element. Consider how you might manipulate the equation to get zero on the right side.

$$3x + -17 + -5y + -13 = 13 + -13$$

$$3x + -17 + -13 + -5y = 0$$

$$3x + -30 + -5y = 0$$

This process avoids the possibility of getting the subtractions confused, doesn't it?

Write out all the steps. You may have noticed that in the problem above, we wrote out all the steps. Most of us naturally like to do steps in our head, but on the exams this leads to errors. The first step above shows adding (–13) to both sides. You could do that in your head. But it is very helpful to get in the habit of writing out every step.

Remember, you are being scored on accuracy, not on brilliance!

Positive and negative values. $x^2 = 9$. What is x? Remember that there are two values: 3 and –3. There are several cases in which you have to be careful about positives and negatives. Squares are always positive, because even with a negative number, the negative times a negative is positive. x^2 is always positive.

Absolute values are always positive. An absolute value, written $|-3|$ (absolute value of –3) can be thought of as the distance that a number is from zero on the number line. That distance is always measured in positive units.

Factoring units. You have seen many story problems that sound like this. John runs 15 miles every day. Every 3 miles he eats 2 apples. In one week John eats 35 apples. For every apple that he eats, John donates $5 to charity. How many dollars does John donate in the month of June?

There is a very easy way to set up this kind of problem so that you can easily avoid the dilemma of what you are supposed to multiply times what. This strategy is called factoring units. Every bit of information in the setup has at least one unit attached to it. Most of the information has two bits of information, such as apples per mile or days per month.

The trick is to set up the information in a way that the all the units cancel out except the ones you want. We can start at the end. What units do we want to end up with? The question asks for dollars per one month. Let's set this up as the final goal on the far right side of our scratch paper.

$$= ? \frac{dollars}{1 month}$$

Next we need to take an inventory of the facts that we know, along with their units. Let's list them across the scratch paper.

$$\frac{15 miles}{1 day} \quad \frac{2 apples}{3 miles} \quad \frac{5 dollars}{1 apple} \quad \frac{30 days}{1 month}$$

Notice that we have included some ratios showing the relationship between days and weeks and days and the one month of June. The trick now is to put these ratios in the correct orientation. 15 miles per day is the same ratio as 1 day per 15 miles. Whether you put it into the calculation right side up or upside depends on what units you want to cancel out. Below is a way to organize all of our information so that everything cancels out except dollars (on top) and 1 month (on the bottom.) Let's start with the ratio that contains dollars. There is only one and it already has dollars on the top.

$$\frac{5 dollars}{1 apple}$$

It has apples on the bottom but we do not want apples in the final answer. Find another ratio that includes apples. Let's use $\frac{2 apples}{3 miles}$. It has apples on the top, which is where it needs to be to cancel out the apples on the bottom in our first ratio.

$$\frac{5 dollars}{1 apple} * \frac{2 apples}{3 miles} = ? \frac{dollars}{1 month}$$

This is the critical point. Do you see how the units "apples" cancel each other out? One on the top times one on the bottom cancels them both. The unit apples is now gone. You can draw a line through them. The **numbers** associated with "apples", though, **must** stay.

$$\frac{5 dollars}{1} * \frac{2}{3 miles} = ? \frac{dollars}{1 month}$$

Now we need to get rid of miles. The only other ratio with miles is $\frac{15 miles}{1 day}$. It has miles on top, where we want it, in order to cancel out 3 miles on the bottom.

$$\frac{5 dollars}{1} * \frac{2}{3 miles} * \frac{15 miles}{1 day} = ? \frac{dollars}{1 month}$$

You continue putting in ratios until you have the units left that you want in the answer. Here is what the whole setup looks like.

$$\frac{5 dollars}{1 apple} * \frac{2 apples}{3 miles} * \frac{15 miles}{1 day} * \frac{30 days}{1 month} = ? \frac{dollars}{1 month}$$

All the units cancel out except dollars on top and 1 month on the bottom. To solve the problem you just multiply and divide the numbers in the order that they appear above. If the units come out right, you have set the problem up right. No need to worry about what gets multiplied by what.

Prime factors. Being able to find the prime factors of a number is an extremely powerful intuitive tool for making otherwise mind-boggling calculations very simple. Consider the problem 1386 divided by 924. One intuitive strategy is just to start finding common factors and reducing. For example both numbers are divisible by 3. However, this can introduce some division errors. By finding the prime factors for each number, the answer is immediately clear. Here is the problem with each number broken into its prime factors.

$$\frac{2*3*3*7*11}{2*2*3*7*11}$$

Everything cancels out except a 3 on the top and a 2 on the bottom = $\frac{3}{2}$.

It is helpful to use an intuitive, visual strategy for finding the prime factors of a number. Here is one way to do that. Let's use 1386. The first step is to use a two-row table. The top row is where you put the factors as you find them. The bottom row is for the numbers you are trying to find factors for. To start out, put 1386 in the bottom left cell.

1386					

Next, you look for prime factors that go into that number, starting with the first prime factor greater than 1. In other words, start with 2. Does 2 go into 1386? Yes. How many times? 693 times. Put the 2 in the cell above 1386. Put the 693 in the cell to the right of 1386 in the bottom row.

2					
1386	693				

Now you simply start the whole process again from the beginning with the number 693. Does 2 go into 693? No. Does 3 go into 693? Yes. How many times? 231. Put the 3 above 693. Put the 231 in the cell to the right of 693.

2	3				
1386	693	231			

Start the process again from the beginning with 231. Does 2 go into it? No. Does 3 go into it? Yes. How many times? 77. Put the 3 above 231. Put the 77 to the right of it.

2	3	3			
1386	693	231	77		

Start from scratch. Does 2 go into 77? No. Does 3 go into 77? No. Does 5 go into 77? No. Does 7 go into 77? Yes. How many times? 11. Put 7 above 77. Put 11 to the right of 77.

Chapter 3. Specific Patterns and Strategies

2	3	3	7		
1386	693	231	77	11	

You can probably see where this is going! Let's finish the process. 2, 3, 5, and 7 do not go into 11. The next prime number is 11. It goes into 11 one time. Put 11 above 11. Put 1 to the right of 11 in the bottom row.

2	3	3	7	11	
1386	693	231	77	11	1

You are done. These are all the prime factors for the number – 2, 3, 3, 7, 11. This process can be used with any number. You need to know at least the first six prime numbers in order to do the testing – 2, 3, 5, 7, 11, 13.

Knowing the prime factors of a number is a powerful tool in many kinds of calculations.

Regrouping. It is really nice on the test when they give you numbers that you can add, subtract, multiply, or divide without having to think. It is completely intuitive to add 10 plus 20 or divide 15 by 3. However, most of the calculations that you have to do are not so intuitive. There are some strategies for converting more complex calculations into an intuitive process. These strategies often involve regrouping the numbers.

Consider the calculation below.

$$13 + 26 + 30 + 7 + 14$$

If you add these from left to right in your head, you might start making mistakes. However, notice what happens if you regroup these numbers.

$$(13 + 7) + (26 + 14) + 30$$
$$(20) + (40) + 30$$

If you were a little shaky on 26 + 14, you could regroup like this.

$$26 + (10 + 4)$$
$$26 + (4 + 10)$$
$$(26 + 4) + 10$$

If you have a clear, intuitive method for keeping track of what you are doing, you can use this technique to add quite complex numbers. Consider 95+84+87, a possible series of scores on a test in a class. We can use a table to break each number into parts that are intuitively easy to add. Here is the step-by-step process. This may be more broken down than you would need to do but this hopefully is helpful for demonstration purposes.

Start by putting the first number, 95, into the top of the left column. To add 84 to it, we can first add 4, which gives a running total of 99 and leaves 80. Then you can use 1 from the 80 to reach a running total of 100, leaving 79 from the original 84.

95	4
99	1
100	79

Now it is easy to add 100 plus 79. Next we can work with the last number, 87. We can use 1 to get to 180 and put the remaining 86 below it.

95	4
99	1
100	79
179	1
180	86

Now you might want to use 80 of the remaining 86 to get to 260. Add on the final 6 for a final total of 266.

95	4
99	1
100	79
179	1
180	80
260	6
266	

The important thing with a strategy like this is that it make sense to you and that it simplifies things for you. It if does not meet those criteria, do not use it. If this specific approach does not click for you, experiment with ways that you can regroup numbers to make calculations easy.

Easier calculations can save you time and can be far more accurate than standard calculation. If nothing else, they can help you quickly double check calculations that you do by standard methods or by calculator.

Averages

There are many questions about averages on the test. First, be clear on the terminology. It is easy to confuse "mean" and "median." The ACT® tells you in the instructions to the math section that "mean" is the same as "average". The SAT® does not tell you that.

Memorize the phrase "Mean means average."

You probably remember that the average of a series of numbers is the sum of those numbers divided by how many numbers there are. If five students have certain scores on a test, you add the scores and divide by 5.

It is sometimes helpful to think of a series of, for example, five numbers with an average of, say, 12, as if each number actually was 12.

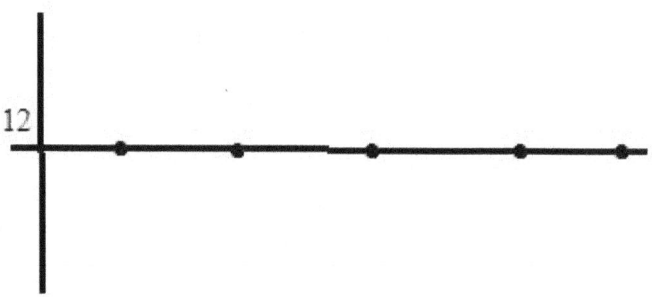

If somehow the first number were changed to 13, what would have to happen to the numbers in order to maintain the average of 12? 13 is 1 unit too high. Do you see that one other number would have to become 1 unit too low?

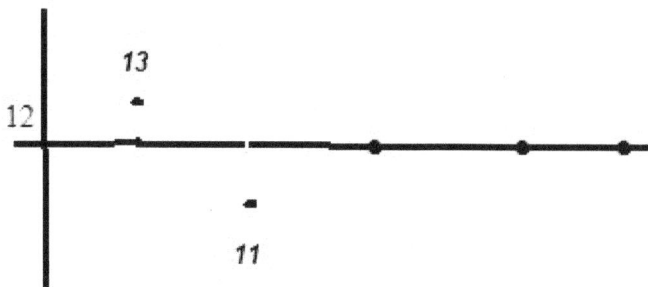

This process and this type of diagram can be extremely helpful in figuring out problems about averages.

Average
"Mean means average."

The median. For a given series, half of the numbers are above the median and half are below. If you have a series with an odd number of members, then the median is one of the members of the series.

$$3, 12, 500, 502, 530$$

The median is 500. Notice that it is nowhere near the average! For a series with an even number of members, the median is half way between the two middle members.

$$3, 12, 500, 502, 530, 1000$$

The median is half way between 500 and 502, so it is 501.

Circles and Triangles

There are some basic intuitive concepts that you need to be clear on for circles and triangles.

Circles – area versus circumference. Many people have a hard time remembering the formulas for area and circumference of a circle. Or they remember the two formulas but do not remember which one is which. There is an intuitive way to remember.

Both formulas include a 2, an *r* and a π (pi). Here they are:

$$2\pi r$$
$$\pi r^2$$

Which is area and which is circumference? Easy. The one with the square is area because area is measured in square units. The other, $2\pi r$, is circumference. But because $2r$ is the same as diameter, circumference can also be written as πd.

Circles – the effect on area of increasing the radius. One trap that the tests set for you is something like this. If a circle has an area of 9π square centimeters and you double the radius, what is the new area? The trap is to say that the area is double, or 18π. Why is that not true?

It is not true because to calculate area, the radius is squared. Let's do the math, first with the original radius of 3 and then with the radius doubled to 6.

$$\pi(3)^2 = 9 \text{ square centimeters}$$
$$\pi(6)^2 = 36 \text{ square centimeters}$$

When the radius doubles, the square of the radius is much more than doubled. This is true not just for circles but for any relationship that involves a square. When a square is involved, doubling is not doubling!

Circles – right cylinder. A right cylinder is a three dimensional object with a circle as its base, just like a box is a three dimensional object with a rectangle as its base. The volume of the cylinder is based on the area of its base times the height of the cylinder.

Chapter 3. Specific Patterns and Strategies

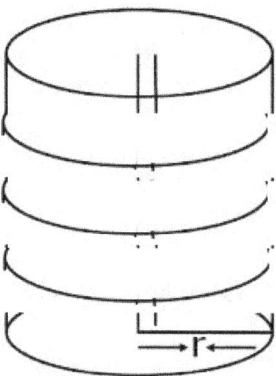

It can help you to visualize a cylinder if you think of it as a stack of pancakes. If the area of the base is, say, 25π square units, then the first pancake is 1 unit high and has a volume of 25π cubic units. If the cylinder is 4 units high, then its total volume is four pancakes or 100π cubic units.

Triangles – area. The area of a triangle is $\frac{1}{2}$ * base * height. If you have trouble remembering that, it may help to understand where the formula comes from.

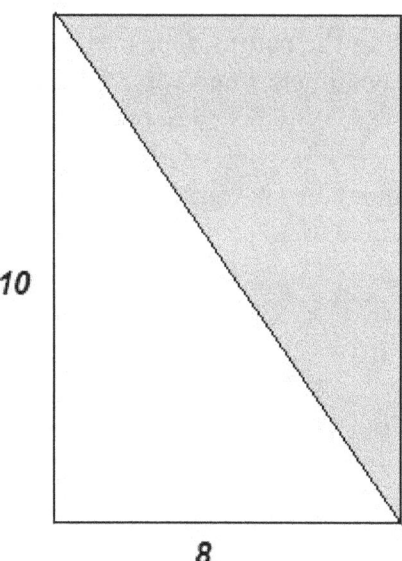

The image above shows a rectangle divided into two equal triangles by a diagonal line. Is it clear that the white triangle is half the area of the rectangle? The area of the rectangle is the base (8) times the height (10). The area of the triangle is one half of that, or $\frac{1}{2}$ bh.

Triangles – isosceles triangle trap. Almost every time you run across an isosceles triangle on the test, it involves the infamous isosceles triangle trap. As you know, an isosceles triangle has two angles that are the same. Because of that, the two sides opposite those angles have the same length. The third angle and third side are different.

The isosceles triangle trap is also known as the teepee trap. Here is an example.

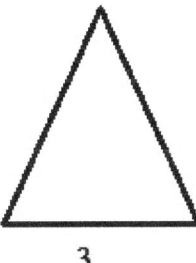

This is an isosceles triangle with a base of 3. The perimeter of the triangle is 10. What are the lengths of the other two sides?

Most people subtract the 3 from 10, get 7 and divide that in half, to make the two "teepee" sides 3.5 each. That's the trap!

The teepee assumption is that when you see the triangle oriented this way, you think of the two sides as having the same length. The base is the odd side out. It makes sense, because you would not want a teepee with uneven sides, right?

Wrong. You cannot assume that the two equal sides are the left and right side. The base might be one of the two equal sides. There are actually two solutions to this triangle and that is what the test wants you to know.

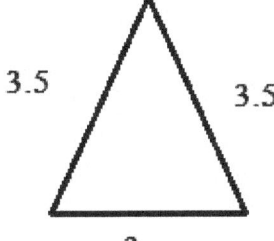 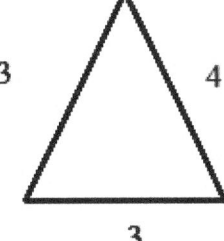

The left triangle is the teepee configuration. In the triangle on the right, the two equal sides are 3, adding up to 6, and the remaining side is 10.

Chapter 3. Specific Patterns and Strategies 27

Triangles – right triangles, Pythagorean Theorem. You are probably at least somewhat familiar with the Pythagorean Theorem, which expresses the relationship between the two sides of a right triangle and the hypotenuse.

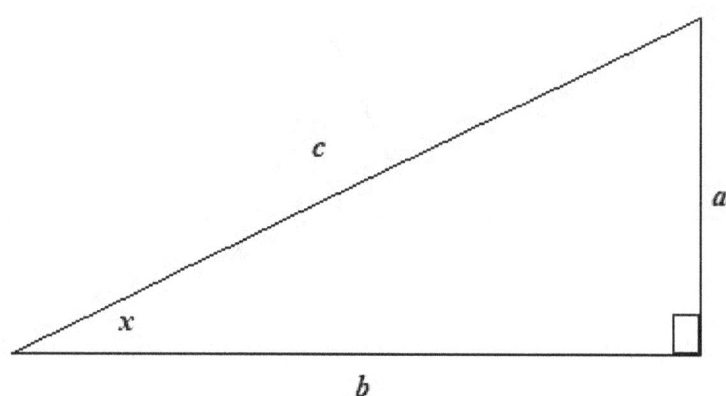

The formula is $a^2 + b^2 = c^2$. The length of the hypotenuse is c, so you have to find the square root of c^2.

Triangles – right triangles, 3:4:5. There are some special right triangles for which you can easily calculate the sides or hypotenuse without having to use the Pythagorean Theorem. A right triangle with sides of 3 and 4 has a hypotenuse of 5. Of course, this is still consistent with the Pythagorean Theorem: 9 + 16 = 25.

There is a 3:4:5 trap, just as there is an isosceles triangle trap.

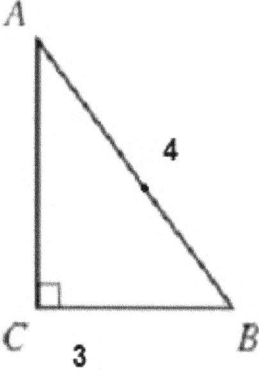

What is the length of the unlabeled side? It is a right triangle. It has a 3 and a 4. Is the remaining side 5? No. Why? In a 3:4:5 triangle the 3 and 4 must be the sides, not the hypotenuse. The test will try to trick you on this.

28   Chapter 3. Specific Patterns and Strategies

Triangles – right triangles, trigonometric relationships. If you are a little shaky on tangents, sines and cosines, not to mention cotangents, secants, and cosecants, here is a brief review. A little memorization work can help you remember these clearly.

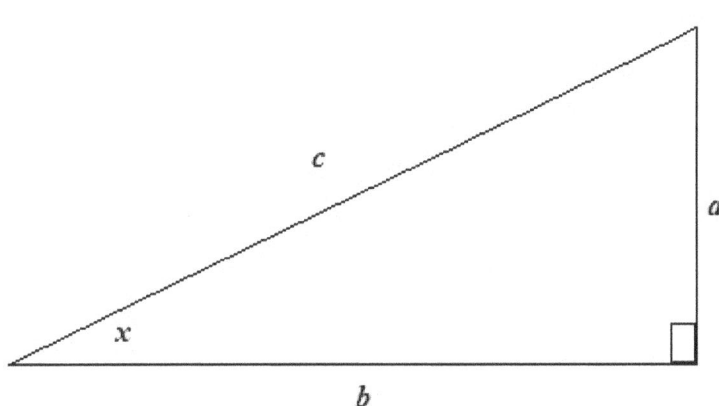

Here is our right triangle again, with angle x. This particular orientation, with the angle on the left, the right angle in the bottom right, is a good standard orientation to use to remember the trigonometric relationships.

Think of yourself as standing at x and looking ahead, out through side a. You have a side in front of you that you are looking at (with length a). You have a floor that you standing on (with length b). Above your head is a sloped roof (with length c).

The tangent, and its reciprocal (upside down) version, the cotangent, are the only two relationships that do not involve the roof (the hypotenuse.) The tangent is the relationship of what you see in front of you over what you are standing on. Look ahead, and then look down. This is a fundamental relationship.

The tangent is usually identified as the opposite side over the adjacent side (or base) - $\frac{a}{b}$. The cotangent is just the reciprocal of the tangent, adjacent (base) over opposite - $\frac{b}{a}$. (Note that there is a term "inverse tangent" that means something different altogether.)

The next most fundamental relationship is the sine. If you think of what you are looking at straight ahead as your home position, the sine is that "opposite" side over the hypotenuse - $\frac{a}{c}$. The cosine is **not** the reciprocal (inverse) of that. The cosine is the adjacent (base) over the hypotenuse - $\frac{b}{c}$.

For sine, cosine, secant, and cosecant you can almost assume that the relationships are the opposite of what you would think. Cosine is not the reciprocal of the sine. Cosecant is not the reciprocal of the cosine.

The reciprocal of the sine is called the cosecant - $\frac{c}{a}$. The reciprocal of the cosine is called the secant - $\frac{c}{b}$.

There is an interesting combination of sine and cosine that you will find on the test. Consider the sine squared added to the cosine squared.

$$(\frac{a}{c})^2 + (\frac{b}{c})^2$$

$$\frac{a^2}{c^2} + \frac{b^2}{c^2}$$

$$\frac{a^2 + b^2}{c^2}$$

According to the Pythagorean Theorem, $a^2 + b^2 = c^2$, so the expression above equals 1 (a number over itself). Thus $\sin^2 + \cos^2 = 1$. You will see this on the test. If you forget that it equals 1, you can figure it out the way we did above.

Triangles – complementary angles. Complementary angles are two angles that add up to 90°. In a right triangle, the two angles that are not 90° must add up to 90° because the total of angles in a triangle is 180°. Angles are also formed when two lines cross.

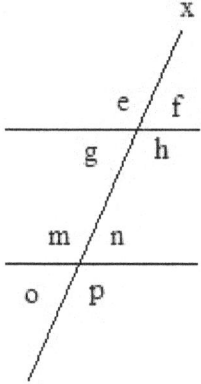

In the image above the two horizontal lines are parallel. They are crossed by line *x*. Notice how, when one line crosses another, there are four angles formed, the four corners of an intersection. Notice that two of the angles are large angles (larger than 90) and two are smaller angles. The two large angles are the same size. The two small angles are the same size.

Look at angles *e* and *f*. Together they form a straight line, which has 180°. A small angle plus a large angle equals 180.

Because the two horizontal lines are parallel, line *x* crosses the lower line at the same angles. That means that large angle *e* is the same size as large angle *m*. You could now prove that angle *m* plus angle *g* add up to 180°, couldn't you, because one large angle and one small angle make 180.

30   Chapter 3. Specific Patterns and Strategies

Ratio, Rate, and Percent

A ratio is a relationship between two elements. Ratios are expressed as "a certain number of something" in relation to "a certain number of something else." This can be expressed as a fraction,

$$\frac{numberOfSomething}{numberOfSomethingElse},$$

or using a colon,

number of something:number of something else.

Some ratios are the relationship between a part and a whole, such as days to week. A day is part of a week. The part is usually expressed first or in the numerator, $\frac{7 days}{1 week}$. Other ratios do not involve a part and a whole, such as *3girls:4boys*, three girls to four boys.

The ratio trap. There is a ratio "trap" that the test sets for you. Consider this problem.

In Elmville 35% of the residents own BMW's. In Tedtown 10% of the residents own BMWs. Which of the following statements must be true?

A. Elmville residents earn more money than Tedtown residents.
B. The BMWs in Tedtown are newer than the BMWs in Elmville.
C. There are more BMWs in Elmville than in Tedtown.
D. There are more American-made cars in Tedtown than in Elmville.
E. It is not possible to know which town has more BMWs.

If you chose C, you fell into the trap. If you chose E, congratulations. The other answer choices are wrong because there is not enough information in the setup for them to be known.

Why is C wrong? Suppose Elmville had 100 residents. That means there are 35 BMWs there. Suppose Tedtown, on the other hand, had 100,000 residents. 10% of that is 10,000 BMWs.

The lesson? Do not confuse ratio with actual number. Would you rather have 50% of a tiny pizza or 10% of a gigantic pizza? Percent (a ratio) does not tell you how much you are getting. It only tells you a relationship between one part and another part.

What ratio actually tells you. If you know a ratio between two elements and if you know the actual number or quantity of one of the elements, you can absolutely calculate the quantity of the other element.

A pickup truck goes 42 miles per hour and travels 14 miles. How long did it take?

Use factoring units to set this up.

$$\frac{1\,hour}{42\,miles} * \frac{14\,miles}{1} = \frac{14\,hours}{42} = \frac{1}{3} hour$$

Notice that in order to end up with hours in the answer, we had to flip 42miles per hour and to remember that the unit "hour" has the quantity 1. For 14 miles, notice that we wrote it as a fraction with 1 in the denominator. The 1 has no units. Similarly, once the miles on the top and the miles on the bottom cancel out, the numbers 42 and 14 no longer have units.

Rates. Rates are specific types of ratios. A rate is the relationship between a certain amount of something to an amount (period) of time, for example 60 miles per hour. This can also be expressed as 60 miles:1 hour or $\frac{60\,miles}{1\,hour}$. Notice that when you say "miles per hour", you really mean "miles per one hour." It is not necessary to write the 1 but it is important to understand that it is there.

Rate problems do not really contain a trap but there is a common type of rate problem that you should be aware of so that you do not go astray.

Ronette makes 30 posters in 2 hours on Tuesday and 36 posters in 3 hours on Wednesday. What is Ronette's rate of production of posters for Tuesday and Wednesday combined?

Rate always consists of total amount over total time. The only way to get an accurate rate for two or more periods is to combine the amounts for both periods and put it over the combined time

$$\frac{Tues\,\&\,Wed\,Amounts}{Tues\,\&\,Weds\,Times}.$$

Many people try to figure out the rate for each day and then add them or average them or do some other random manipulation. Go right to the totals.

Rate for Two or More Periods

$$\frac{TotalAmountForAllPeriod}{TotalTimeForAllPeriods}$$

Consider this problem.

Edna runs 15 miles per hour. Ed runs 10 miles per hour. On Saturday both Edna and Ed ran along the river. Who ran further?

Ratio is not the same as actual number. Edna's ratio (rate) is higher, so Edna ran faster, but we do not have enough information to know who ran further. We would need to know how long each person ran in order to know how far they ran. Do not confuse rate with actual number (quantity).

Percent. Like rate, percent is a specific type of ratio. If 1 person out of 5 has red hair, how many people will have red hair in a crowd of 100 people? A percent is a ratio expressed in terms of 100 elements (people, parts, pieces, etc.) 1 out of 5 is the same ratio (fraction) as 20 out of 100.

$$\frac{1}{5} = \frac{20}{100}$$

Note that the word "percent" literally means "out of" (per) one hundred (cent). You can think of the word "per" as being represented by the division bar. $\frac{20}{100}$ is (20) (out of= per=−) (100=cent).

If these are the same, why not just reduce the fraction as far as possible and express the relationship as 1 out of 5? There are certain advantages to using percent. Percentages can be compared with each other more easily than comparing a variety of fractions. If 1 out of 5 people has red hair and 3 out of 7 people have black hair, how many people in a group of 50 have either red or black hair? Too confusing. But if 20 percent have red and 40 percent have black, the total is 60 percent.

Many problems require that you change a ratio to a percent. For example, if 6 people in the Geometry class have black hair and there are 15 people in the class, what percent of the class have black hair? You can express this problem as below.

$$\frac{6}{15} = \frac{?}{100}$$

Of course it is possible to solve this with algebra, but usually that is the least accurate method. As you go through the problems in this book, you will learn a variety of intuitive ways to get the answer to a problem like this. For example, notice that $\frac{6}{15}$ is a little less than half. That means the answer must be somewhat less than 50. You can often work backwards from the answer choices. In this case any answer that is 50 or more is out.

Ratios as building blocks. There is one important tool for working with ratios that we will look at here. As you go through problems, you will learn about this tool in more depth.

As you know, a ratio can be expressed in many – actually, an infinite number of – ways. Suppose that 27 people out of 60 really like you. Suppose you want to know how many people out of a hundred really likes you and you also want to know many people in your History class of 23 people really like you.

The intuitive strategy below allows you to break ratios into small building blocks that you can use to create new expressions of the same ratio. For example, if you knew a certain ratio was 0.31 friends out of every 1 person, you could figure out how many friends out of 2 people just by doubling 0.31. You could find out how many friends out of 3 people by tripling 0.31 and so on. But that process would quickly become cumbersome. By creating more workable building blocks, you can easily find equivalents.

Let's try to figure out how many people out of 100 really like you and how many out of your class of 23. Take the ratio 27 people who really like you out of every 60. Put it into a table. For every 60 people there are 27 who really like you. In the next row, let's create a smaller version of the ratio to make a more flexible building block. We can divide both 60 and 27 by 3. Enter 20 and 9 in the second row.

60	27
20	9

Here is the power of a building block. If you want to know how many people really like you out of 80, you could combine the two rows! That is how the building blocks work. If you want to know 40, you could subtract the row for 20 from the row for 60. However, subtraction is always more prone to error than addition. We'll stick with addition. Next, double the row for 20. You will see why in a minute.

60	27
20	9
40	18

You can now calculate the value for 100 by adding the rows for 60 and 40. It comes out to 45 people out of 100 who really like you, or 45%.

There are many different combinations that can get you to your answer. You could have added the rows for 60 and 20 and then added the row for 20 to that result.

You still want to know how many people really like you in your History class of 23. We have a row for 20 people. How can we get a row for 3? Often you do not know in advance how you will get there but we can try creating some smaller building blocks. Divide the row for 60 by 10. Enter 6 and 2.7. Now you can divide that row in half to get 3 and 1.35. Finally, add the rows for 20 and 3. 23 = 10.35.

60	27
20	9
40	18
6	2.7
3	1.35
23	10.35

If this were an actual test question, you probably would not have had to work all the way to the final answer. You probably could have checked the answer choices and found the correct answer through elimination of choices that were too high or too low.

In using this method, the most foolproof steps are often dividing by 2 or 3 or by 10. Larger building blocks help you get to higher numbers. Smaller building blocks help you narrow it down to more precise answers.

Exponents, Roots, and Logs

Exponents, roots, and logs are related to each other. They express the same relationships in different ways.

$$5^3 = 5 * 5 * 5$$
$$\sqrt[3]{5*5*5} = 5$$
$$\log_5 5*5*5 = 3$$

The statements above are not necessarily easy to understand. Throughout this book we will break this down and look at various intuitive strategies for keeping these relationships straight.

Exponents. Most likely you are comfortable with whole number exponents. $5^3 = 5 * 5 * 5$. It is helpful to think of a number raised to a certain power as a string of beads. The 5 is the bead. In other words, each bead is a 5. In this string there are three beads. Beads are connected by multiplication. The multiplication is like the little knob that connects one wooden bead to another.

This concept is helpful if you have to multiply 5^3 times, say, 2^5. The bead concept makes this easy.

$$5*5*5*2*2*2*2$$
$$(5*2)(5*2)(5*2)*2*2 - (\text{regroup})$$
$$10*10*10*4$$
$$1000*4$$
$$4000$$

The exponents that give most people a little bit more of a challenge are fractional and negative exponents.

$$(5*5*5)^{-3} = \frac{1}{5*5*5}$$

$$(5*5*5)^{\frac{1}{3}} = \sqrt[3]{5*5*5}$$

How can you remember what a negative and fractional exponent mean? There are just two things to keep straight. The negative exponent and the fractional exponent really mean a fractional expression and root, respectively. One way to remember this is that the fractional exponent does **not** lead to a fractional answer.

There is actually a way that you can derive the reason that a negative exponent leads to the fraction but it is probably easier just to memorize that the fractional exponent does **not** lead to the fractional answer.

Fractional and Negative Exponents	
Exponent	**Result**
negative	fraction
fraction	root

Chapter 3. Specific Patterns and Strategies

Roots. Most roots on the test are square roots. There are some ways in which roots become confusing. It is probably clear to you that the square root of 4 is 2, the square root of 9 is 3, the square root of 16 is 4, and the square root of 25 is 5.

What may be less clear is that the square root of 5*5 is 5. The square root of 4*4 is 4. It may help to think of a square root as a sort of "half." To find half of, say, 12 apples, you divide the apples into two equal piles. Adding the two piles together gives you 12 and each pile is exactly the same size. That is the normal concept of half.

A square root also divides a number in half but not in terms of adding. In terms of multiplying. Two "halves", when multiplied together, give you the original number. To get the square root of 25, you have to determine that 5*5 gives you 25.

You can approximate a square root just by trial and error. What is the square root 53? 6*6 is 36. 7*7 is 49. That's closer. 8*8 is 64. That's too much, so the square root of 53 is between 7 and 8.

To find the square root of a more complex number, it often helps to break the number into factors, usually prime factors. Let's find the square root of 1764 by breaking it down into prime factors. We can use the intuitive process for finding factors.

2	2	3	3	7	7	
1764	882	441	147	49	7	1

The factors are 2*2*3*3*7*7. What is the square root of that? Can you divide that sequence into two exactly equal parts?

$$2*2*3*3*7*7$$
$$(2*3*7)(2*3*7) - \text{(regroup)}$$
$$\sqrt{2*2*3*3*7*7} = 2*3*7$$

This process reminds some people of the process of mitosis in biology. You can also think of it as divvying up a set of things into two equal piles. Each pile gets a 2. Each pile gets a 3. Each pile gets a 7. When you see an expression like $\sqrt{2*2*3*3*7*7}$, remember that you are just supposed to take that long string and divide it into two equal piles.

How would you do this problem - $\sqrt{2*2*3*3*5}$? You can put a 2 and a 3 in each pile but what about the five? If you could divide the 5 into two equal parts, you could easily make two exact piles. In fact, you **can** do that.

$$\sqrt{2*2*3*3*\sqrt{5}*\sqrt{5}}$$

Now each pile contains $2*3*\sqrt{5}$.

Logs. Logs are just another way of expressing a relationship with exponents. Consider $5^3 = 125$. Using the "bead" concept, this can be written as:

$$5*5*5 = 125$$

There are three components of this relationship, right? There is the bead, which is a 5. We can call this the "base" because it is what the string of beads is composed of. There is the result (125). There is the number of beads in the string.

$$\text{Base}^{numberOfBeads} = \text{Result}$$

Another way to express this relationship would be to say "For a certain base, if the result is Result, what is the number of beads?" If you can memorize this, you will understand logs. Let's simplify it.

Here's the base. Here's the Result. What is the number of beads?

$$\text{Log}_{Base} \text{ Result} = \text{\# of beads}$$
$$\text{Example: Log}_5 125 = 3$$

When the bead (base) is 5 and the result is 125, there have to be 3 beads.

It helps to remember that, when a questions ask you what the log is, it is asking what the exponent is. They tell you the base. They tell you the result. You have to figure out the exponent that makes it work. That's the log.

Hopefully, this gives you some intuitive tools for keeping logs straight. It is not usually that you do not understand logs. The problem is that it is easy, especially in the middle of a test, to get the elements confused. Intuitive tools help you stay accurate with information that you may already know but could get confused on. And the test writers **are** trying to confuse you.

Gaining points on the test often has more to do with being highly accurate than with your overall knowledge. Intuitive tools keep you accurate.

Equations and Inequalities

There are just a couple basic points about equations and inequalities that you need to know now. You will learn more as you work through the problems in this book.

Equations. An equation with one variable can be solved. $x + 3 = 10$. $x = 7$. In some cases there is more than one solution. $x^2 = 25$. $x = 5, -5$.

One equation with two variables cannot be solved. There are many combinations of x and y that make the equation work. However, if you have two equations with two variables, you can solve for x and y. There is one trick to this. The two equations have to be different. $x + y = 3$ and $2x + 2y = 6$ are the same equation, just in different forms.

There is one important intuitive tool that you can use to solve a system of two equations in two variables. You have probably learned this.

$$x + 2y = 10$$
$$3x + 4y = 12$$

The trick is to add or subtract the equations so that one variable cancels out. Then you have an equation in one variable that you can solve. Once you determine the value of that variable, you can use that information to calculate the other variable. In the above system, you need to write one of the equations in an equivalent form. If you decide to cancel out the *x*'s, you could rewrite the first equation to get either $3x$ or $-3x$. If you go with $3x$, you would subtract the equations so that the x factors disappear. If you go with $-3x$, you can add the equations to make *x* disappear.

$$3(x + 2y) = 3 * 10$$

$$3x + 6y = 30$$
$$-\ (3x + 4y = 12)$$

$$2y = 18$$
$$y = 9$$

Inequalities. There is just one fact about inequalities that can trip you up and the test will try to catch you on it. You can add or subtract any amount from both sides of an inequality, just as you can with an equation.

For multiplication and division, you can do the same as long as the number is positive. If you multiply or divide by a negative number, you must reverse the inequality signs.

When you have a three part inequality, such as $2<x<7$, it is often more intuitive to break this into two separate statements. $2< x$ and $x<7$.

Finally, sometimes you need to switch the direction of an inequality statement, such as changing $2<x$ to $x>2$. Remember that it is very easy to make a mistake when you do that. Double-check yourself!

Quadratic equations. Generally, quadratic equations on the test that needs to be factored, it can be done so with relatively simple numbers. You should make yourself familiar with the process of factoring a quadratic.

$$x^2 + 2x - 15 = 0$$
$$(x + a)(x + b) = 0$$

At this point you have to find *a* and *b*. This usually takes some trial and error but on the test the numbers will be relatively straightforward. You know that $a + b$ has to equal $+2$. You also know that ab must equal -15. The most intuitive way to find the numbers is to list the pairs of factors for 15. They are (1,15) and (3,5). Only the second pair has a difference of 2.

Now you have to be really careful to make sure that the signs (positive or negative) are correct.

$$(x + 5)(x + -3) = 0$$

To double-check yourself, FOIL the expression. Once you have the correct numbers, there is one more trap you have to avoid. Usually you have to find the values for *x*. The values for *a* and *b* are 5 and –3 but these are **not** the values for *x*.

Because $(x + 5)(x + -3) = 0$, either $(x + 5)$ is zero or $(x + -3)$ is zero. If two numbers are multiplied to get zero, one of them has to be zero.

$$(x + 5) = 0$$
$$x = -5$$
or
$$(x + -3) = 0$$
$$x = 3$$

The values of *x* are –5 and +3.

$m^2 - n^2$. This is a special relationship that you will run into on the test. For example, what is $83^2 - 17^2$?

If you were to do the actual calculations, there is a good chance that you would make a calculation error or would have to spend a lot of time checking yourself. The pattern you need to know is:

$$m^2 - n^2 = (m+n)(m-n)$$

$$83^2 - 17^2$$
$$= (83+17)(83-17)$$
$$(100)(66) = 6600$$

You can prove why this works by FOILing $(m+n)(m-n)$. Some other patterns that can be helpful to recognize are $(m+n)(m+n)$, which is $(m+n)^2$, and $(m-n)(m-n)$, which is $(m-n)^2$. FOIL both of these for yourself so you can see what the results look like.

x,y Coordinates

You will find many questions that draw on an (*x,y*) coordinate system, mostly involving equations for lines but occasionally involving sine waves, points, or other figures.

Slope. Many questions involve the slope of a line. You may remember that the slope is defined as the rise over the run. As you go from left to right in the figure below, starting at point A, the run is the distance that you go measured along the *x*-axis. Going 2 units to the right takes you 1 unit up (the rise) measured on the *y*-axis.

The rise over the run in this case is 1:2. Typically you count the run first, so you have to remember which number is rise and which is run. We will look at intuitive strategies for working with slope in a number of problems in this book.

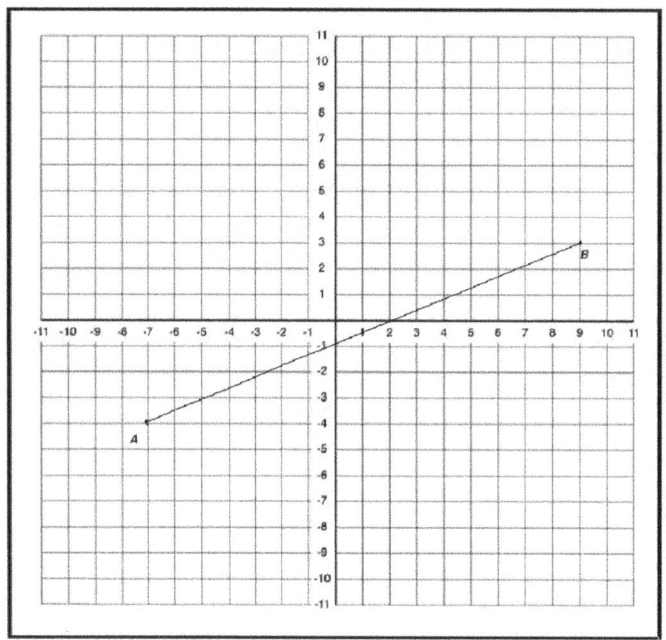

Finding points on a line. If you are given the formula for a line, the most intuitive way to find points is to set x to zero in the equation and find the value for y. This gives you a point $(0,y)$. When x is zero, the point is on the y-axis. Similarly, you can set y to zero and solve for x. This is the point $(x,0)$, which is on the x-axis.

f(x). The expression $f(x)$ refers to a function of x and this simply means that the expression is defining a process that you are to perform on any number. In real world terms, we could define $f(x)$ as meaning when you meet person x, you shake their hand and move them over 3 feet to the right. This is a process. f(Frank) means you shake Frank's hand and move him over 3 feet to the right.

In the context of an (x,y) coordinate system, $f(x)$ usually is synonymous with y. The two equations below represent the same thing. You plot $f(x)$ as the y coordinate.

$$y = 3x + 10$$
$$f(x) = 3x + 10$$

Summary

Reviewing the strategies and patterns presented in this chapter will help you get more out of the explanations in this book. They also give you a synopsis of some of the main patterns that you will find on your test. If you want to get a more extensive preview of patterns on the test, take a look through the index at the back of the book. The index also helps you find specific questions in the book that cover a certain topic or pattern.

The best way to learn these patterns and strategies is through hands-on work. As you go through the problems in the book, try them on your own, and then read the explanations thoroughly. Everything in this chapter, plus much more, is covered in great depth in the explanations. Once you have gotten what you can out of this chapter, move on to the test questions in the next three chapters and dive in to the world of intuitive math.

Chapter 4. ACT®-style Questions with Explanations

The questions in this chapter are written and formatted in the style of questions on the ACT® exam. They represent the most common patterns of math questions found on both the ACT® and SAT® exams. On ACT® questions you are allowed to use a calculator. However, using a calculator is not always the most accurate way to get an answer.

Try to work each problem on your own before reading the explanation. Take as much time as you need. It can be helpful to give yourself 15 or even 30 minutes or more to work on a problem. The longer you work on it, the more you can learn.

The explanations are designed to be simple and intuitive. Nevertheless, you will probably find some explanations challenging. Stick with it. Experiment with it. Work on it with a friend.

In some problems you might find that the intuitive strategies seem unnecessary and that the problem can be easily solved with standard math tools. We suggest that you still study the intuitive tools for that problem. You may need those tools on another problem.

Your primary focus is to learn new ways of thinking about mathematical relationships. It does not matter that much whether you get a question right or wrong as you practice. The patterns in this chapter are the patterns you will see on your test. Study them, learn them, and get comfortable with intuitive tools for solving them.

You can follow the daily assignments – about five questions per day – or you can do more or fewer questions per day. At the end of each day's assignment you can evaluate how you did.

Day 1, Questions 1-5

1. Which pair of equations below has the same solutions as the equation $|7y - 3| - 3 = x$?

A. $7y - 6 = x$
       $-7y - 6 = -x$

B. $7y - 3 = x + 3$
       $-7y - 3 = x + 3$

C. $y = \dfrac{x+6}{7}$
       $y = \dfrac{x-6}{7}$

D. $7y - 3 - 3 = x$
       $-(7y - 3) - 3 = x$

E. $7y - 3 = x + 3$
       $7y = x + 6$

Explanation of 1:

Orient yourself carefully to the question stem. Many wrong answers are simply due to the fact that you may not have really understood what the question is asking!

In this question you are being asked to find a pair of equations that has the same solutions as the original equation.

For an equation with one variable, such as $3x = 12$, it is possible to find one value for x that makes this work (in this case $x = 4$). For an equation with two variables, such as $y = 3x$, there may be an infinite number of x,y pairs that satisfy the equation, such as (1,3), (2,6), (5,15) and so on.

It helps to know that you will not be able to find exact numbers for x and y. So what does it mean to find a set of equations in the answer choices that will have the same solution as the original equation? Most likely this means that the correct answer choice is simply a rewriting of the original equation, just as $2y = 6x$, is really just the same equation as $y = 3x$.

The fact that you have to find an answer choice with two equations, not just a rewrite of the original equation, is interesting. Ask yourself how the original equation might result in two distinct equations.

Do you see that it has to do with the absolute value? In case you have forgotten, the absolute value of a number (represented as $|x|$) is the distance that the number is from zero on the number line. This distance is always expressed as a positive number! So –3 and 3 are both 3 units from zero.

$$|3| = 3$$
$$|-3| = 3$$

Another way to express this is that $|x| = 3$ is the same as saying:

$$(x) = 3 \text{ and } -(x) = 3$$

Do you see how you can apply this to the original equation in the question to create two different expressions?

$$(7y - 3) - 3 = x$$
$$-(7y - 3) - 3 = x$$

At this point you could be tempted to do some simplification and rewriting of these two equations. Remember the principle "Check to see if I already have enough information to answer the question"! You do. Choice D is essentially what you have already figured out. You're done! The lesson here is to avoid doing more work than you need to. Check regularly to see if you already have enough information to answer the question.

2. For the function f(x) = sin(4x) + 5, which of the following represents the period?

F. 2π + 5

G. 2π

H. π + 5

J. $\frac{\pi}{2}$

K. $\frac{\pi}{4}$

Explanation of 2:

If you find this content challenging, you're not alone. We'll review the concepts that you need to deal with graphs of trigonometric functions.

A sine function repeats a certain pattern over and over. It starts at what we can call zero, goes through a series of phases, and returns to zero – just like phases of the moon or hours of the day or the wheel on a bicycle! The "time" that it takes to go through the entire cycle and return to the starting point is called the period.

Here is a picture of one cycle of sine wave.

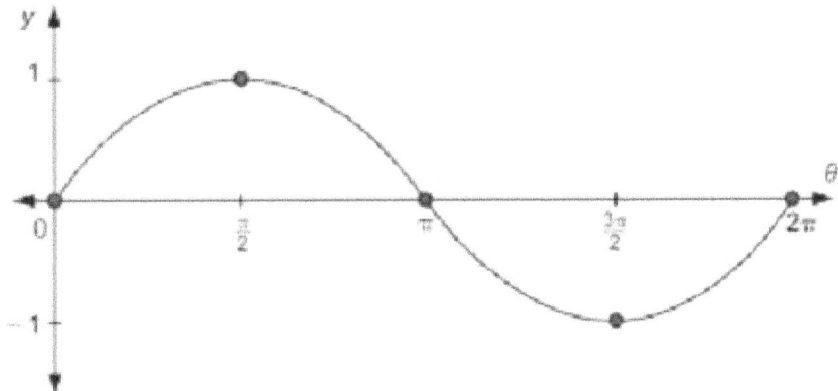

On the left is the starting point. Find the ending points. Do you see the halfway point? And for each half can you identify the halfway point of that half (one quarter of the cycle)?

The cycle starts at zero. The first half of the cycle is in the positive area (above the *x* axis), meaning that all of the *y* values are positive. If we describe the cycle as y = *sin* x, that means that for the first half of the cycle *y* is positive.

However, notice that in the first quarter of the cycle, the value of y is increasing, whereas in the second quarter y is decreasing (and yet still positive).

In the second half of the cycle y is always negative. However, it becomes increasingly negative in the third quarter and starts to increase (though still negative) in the fourth quarter. The table below summarizes the cycle.

1st quarter	Positive	Increasing
2nd quarter	Positive	Decreasing
3rd quarter	Negative	Decreasing
4th quarter	Negative	Increasing

It's like a roller coaster that turns upside down half way through!

Notice on the image above that each quarter is labeled with a value of π. The first quarter is $\frac{\pi}{2}$. The second is π. The third is $\frac{3\pi}{2}$ and the fourth is 2π. For right now you don't need to worry too much about how π gets involved in this. It has something to do with circles and the fact that a sine wave basically goes in a circle (starts at one point, completes a cycle, and returns to the beginning.)

This "periodicity" of the sine function (one cycle is one period, which repeats) is sometimes represented on a circle as below.

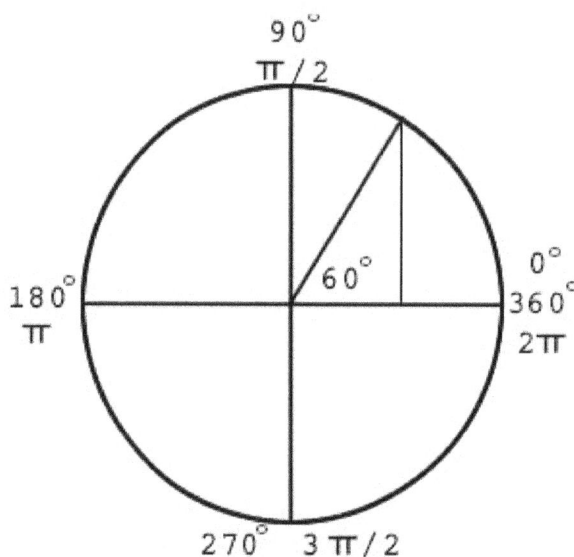

Let's consider how this relates to the sine as a function of a triangle. The picture shows us what the sine of 60° looks like when graphed in the circle. It falls in the first quarter of the cycle (Quadrant I) and is thus a positive number. Notice that the value of sine 60° falls between 0 and $\frac{\pi}{2}$.

44 Chapter 4. ACT®-style Questions with Explanations

Let's see if this is enough background to go back and try to understand the problem. In the picture of the sine function above, the period is 2π because 2π is where the cycle begins again. This picture is the graph of $f(x) = \sin x$.

The test will try to confuse you by modifying this equation in one or both of two ways.

$$f(x) = \sin x + 3$$
$$f(x) = \sin 3x$$
$$f(x) = \sin 3x + 3$$

In other words they can add a numerical amount to the equation, as in the first example. They can multiply x by a numerical amount, as in the second example. They can do both, as in the third example and as in this question.

Take a deep breath. We're going to look at what each of these two manipulations does to the graph, so that you can understand them.

Consider the basic equation $f(x) = \sin x$. Let's look at some points on that graph.

$$(0,0), (\tfrac{\pi}{2},1), (\pi,0), (\tfrac{3\pi}{2},-1) \ (0,0) = (2\pi,0)$$

The table below shows that for the equation $f(x) = \sin x + 3$, for each value of x, we just add 3 to the original value of y.

Value of x	$f(x) = \sin x$ (value of y)	$f(x) = \sin x + 3$
0	0	3
$\dfrac{\pi}{2}$	1	4
π	0	3
$\dfrac{3\pi}{2}$	-1	2
2π, 0	0	3

This gives us five new points:

$$(0,3), (\tfrac{\pi}{2},4), (\pi,3), (\tfrac{3\pi}{2},2) \ (0,3) = (2\pi,3)$$

Below is what that graph looks like. Notice that the graph has simply risen three units. Its shape is the same. Its period is the same. You've simply added 3 to all of the y values of the original equation.

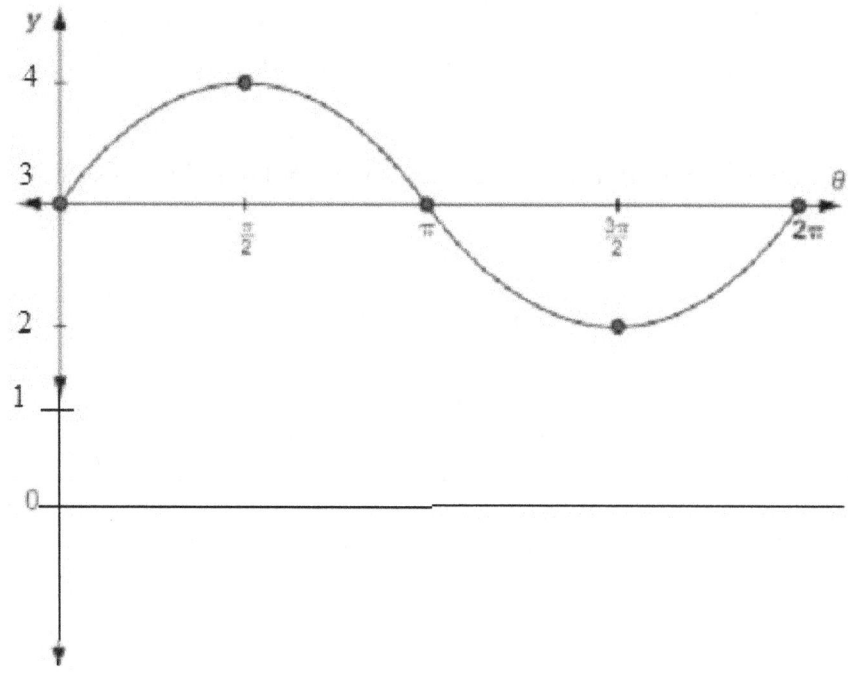

If this makes some sense to you, you can simply memorize that the equation f(x) = sin x + a just raises or lowers the graph, depending on whether a is positive or negative.

Consider the original equation in this problem: f(x) = sin(4x) + 5. Now you know that adding 5 will not change the shape of the graph or the period of it. It simply raises the graph up 5 units on the y-axis.

But what does multiplying x by 4 do to the graph? This is not super easy to understand. To get a grip on it, let's consider a simpler example, f(x) = sin(2x).

Here's one explanation that may help. Let's suppose that you want the new equation to act like the basic sine wave, f(x) = sin (x).

Value of x	f(x) = sin x (value of y)
0	0
$\frac{\pi}{2}$	1
π	0
$\frac{3\pi}{2}$	-1
2π, 0	0

We can redo the chart and just call the first column 2x, instead of x. In a way we are just giving the x value a new name. We could call it z or ♣ or Ed. We will call it 2x because that tells us the relationship between our new x variable and the original one.

Value of 2x	f(x) = sin x (value of y)
0	0
$\dfrac{\pi}{2}$	1
π	0
$\dfrac{3\pi}{2}$	-1
2π, 0	0

The new *z* or ♣ or Ed acts just like our basic sine wave, right? Now we can figure out what the *x* value on the actual *x*-axis would have to be to give us the Ed we are looking for at each point. We do that by dividing the value in Column 1 by 2.

Value of 2x	Actual x value ($\dfrac{1}{2}$ Col 1)	f(x) = sin x (value of y)
0	0	0
$\dfrac{\pi}{2}$	$\dfrac{\pi}{4}$	1
π	$\dfrac{\pi}{2}$	0
$\dfrac{3\pi}{2}$	$\dfrac{3\pi}{4}$	-1
2π, 0	π	0

What happens if we plot these values, using the second column as the value for *x* and the third column as the value for *y*?

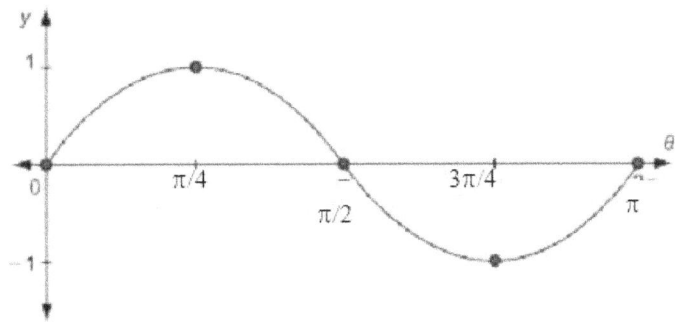

This looks like the basic sine wave but notice that it ends its cycle at π, not at 2π. We have scrunched the wave in half along the *x*-axis.

This may not be crystal clear to you but you can memorize that for a function f(x) = sin (kx), the period of the basic sine wave (2π) will be divided by k to $\frac{2\pi}{k}$.

Given that, for this problem k = 4, so the new period is $\frac{2\pi}{4}$, or $\frac{\pi}{2}$. Choice J is correct.

Trigonometric relationships are not very intuitive. If you play with them a bit, you may find that helps build your intuitions.

For f(x) = sin (kx) + a, it might help to think of it as f(x) = sin ((left/right)x) + (up/down) because the k scrunches the wave left or right and the *a* pushes the curve up or down.

There are other factors that stretch the curve up and down but you don't need to worry about those now!

3. A school newspaper conducted a survey in which they asked every one of the 250 members of the junior class whether they had ever played chess or basketball. The newspaper printed the following results.

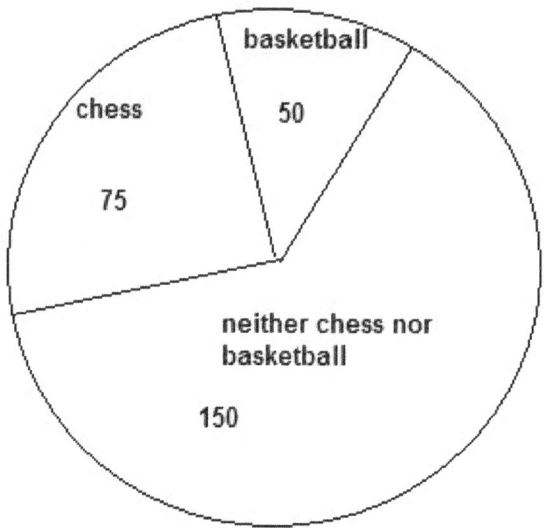

One student, May, pointed out that the chart contained two errors. One error was that the chart implied that there was no overlap between people who have played chess and people who have played basketball. If the numbers in the chart are correct and if all students responded to the survey, what is the number of students who have played both chess and basketball?

A. 5
B. 10
C. 25
D. 50
E. 100

Explanation of 3:

A circle graph should represent the entirety of something. Orient yourself to what the circle graph is supposed to represent. It is the junior class, which consists of 250 students. Now check to see if the circle represents 250 students.

It doesn't. The numbers, as printed in the newspaper, add up to 275. You have discovered one of the errors in the graph. Assuming the numbers are correct, there must be an overlap of 25 students who played both chess and basketball in order to bring the total to 250. The correct answer is C.

Another way to look at this is that the area containing chess and basketball players must represent 100 students. The only way to draw that is to have an overlap area of 25.

---

4. The top scores on a final exam in History are shown below.

Student	Score
M	82
N	88
O	85
P	88
Q	93
R	94
S	85
T	90
U	94
V	97

What is the median score for the data in the table?

F. 93.5
G. 90
H. 89.6
J. 89
K. 88

Explanation of 4:

Orient yourself carefully to the question stem. The word "median" is tricky. The median is NOT the average. "Mean" means average. It's very easy to get these two terms confused during the test, even if you are clear on them outside of the test.

The median is the number for which half of the elements are above it and half are below. To find the mean you have to put the scores in order. Don't be misled by the order that they are shown in the table. Also notice that the names of students associated with the scores are irrelevant. The answer does not require you to identify a student.

The numbers in ascending order are shown in the table below.

Score
82
85
85
88
88
90
93
94
94
97

To find the median, first count how many elements there are. There are 10. That means there will be five elements above the median and five below. The median is therefore not one of the numbers in the chart. It is between two of them.

Find the lowest five elements. They are 82, 85, 85, 88, and 88. The five highest elements start with 90. Therefore the median is half way between 88 and 90. The median is 89, as in choice J.

5. Comet A is visible from Earth every 9 years. Comet B is visible from Earth every 15 years. If at a certain instant both comets are visible from Earth at the same time, how many years will it be before they are next visible from Earth at the same time?

A. 9
B. 12
C. 15
D. 24
E. 45

Explanation of 5:

In theory this can be solved by an algebraic equation but that is prone to lots of error and is not at all intuitive. Let's organize the information visually. Starting from the point at which both comets are visible, we can simply write down for each when it will next be visible and then look for a time when they are both visible.

Comet A	Comet B
9	
	15
18	
27	
	30
36	
45	45
	60
	75

Notice that the table is "to scale". That makes it easy to spot the number that is the same in both columns. You have your answer, which is choice E, 45.

In essence you need a number that is both a multiple of 9 and a multiple of 15. To find the lowest such number you could break both numbers into their prime factors.

$$9 = 3 * 3$$
$$15 = 3 * 5$$

The lowest common number would have to have factors that include two 3's and one 5:

$$3 * 3 * 5 = 45$$

You may notice that using the table is more visual and more intuitive than looking for the common factors. It's a more foolproof way of solving this problem. However, working with prime factors can also be a very powerful tool on problems that can't be organized visually.

You finished Day 1! How did it go on these five questions?

Number of questions you got right on your own: _____

Types of problems or patterns you need more work on: _____

_____

How much new did you learn from these questions? ☐ Important tools! ☐ Some tools ☐ Not too much

Day 2, Questions 6-10

6. Andre wants to wrap a present in a right cylindrical box. The base of the box has a diameter of 10 inches. The box is 15 inches long. What is the total area in square inches of wrapping paper that Andre must use to wrap the entire box, assuming that there is no overlap of paper?

(Note: The surface area of a cylinder consists of the surface area of the two circular ends, represented by $2\pi r^2$, plus the surface area of the side, represented by $2\pi r l$, where l is the length of the cylinder and r is the radius of the circular base.)

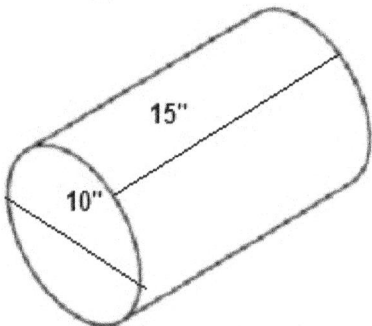

F. 15π
G. 50π
H. 200π
J. 225π
K. 1500π

Explanation of 6:

This question looks harder than it is. The setup actually gives you most of the relationships you need. The test often tries to intimidate you with complex-looking questions. Don't be fooled!

Orient carefully to the situation. Do you see that the question just wants you to calculate the surface area of the cylinder? The wrapping paper itself is irrelevant. The note after the question stem helps you orient to the fact that the total surface area equals the area of each of the two ends plus the area of the remaining part of the cylinder. The note also gives you the formula for calculating these. Theoretically, you just have to plug the numbers into the formulas. However, it may be helpful to remind yourself where these formulas come from.

Each end is a circle. The area of a circle is πr^2. There are two ends so the combined area of both ends is $2\pi r^2$.

[If you are fuzzy on the area of a circle, here's a hint. There are two measures of a circle that you are commonly tested on – the area and the circumference. Both measures include the characters π, r and 2. That's why many people get them confused.

The two formulas are πr^2 and $2\pi r$. Which one is area and which one is perimeter? Area is measured in square units, so the formula with the square is the one for area.]

What about the remaining part of the cylinder? It looks round but do you notice that it is simply a rectangle folded around on itself? Its area will be its length, *l*, times its width, which is the perimeter of the circle on the end ($2\pi r$).

There are a number of steps to the calculations for this problem, so let's create a road map. Remember that creating a road map allows you to focus on the big picture and then write it down. That way, when you start getting bogged down in the details, you'll be able to look back at the big picture.

>   Step 1. Calculate the area of one end, using πr^2.
>   Step 2. Double the number in step 1, to include both ends.
>   Step 3. Calculate the area of the rest of the cylinder, using $2\pi rl$.
>   Step 4. Add steps 2 and 3.

Step 1. The radius of the circle is 5 (half of the diameter of 10). $\pi r^2 = 25\pi$
Step 2. Doubling step 1 gives 50π.
Step 3. $2*\pi*5*15 = 2*5*15\pi = 10*15\pi = 150\pi$
Step 4. Step 2 plus step 3 = 200π.

The answer is choice H.

Notice that in step 3 we avoided having to do a complex multiplication by regrouping the numbers. Multiplying 5 times 15 is more complex and more prone to error than multiplying 2 times 5. And then it is intuitively easy to multiply 10 times 15.

---

7. The average of four numbers is 8. The median of the same four numbers is 7. Which of the following could be the four numbers?

A. 3, 5, 9, 11
B. 6, 6, 8, 12
C. 5, 6, 7, 14
D. 6, 7, 7, 8
E. 6, 7, 7, 10

Explanation of 7:

This question combines the oft-confused concepts of mean and median. On the ACT® the directions actually define average as arithmetic mean. The SAT® does not do this.

Let's review averages. Consider five numbers that average 12. In certain ways these five numbers act as though each of them were 12. We can plot these numbers on a number line.

The vertical line on the left indicates that the horizontal line is at 12 units. Because each point is exactly 12, the average of the five numbers is 12. Consider what would happen if the fourth point were NOT at 12. Let's say it is 15. In order to keep the average 12, you would have to change one of the other points, wouldn't you?

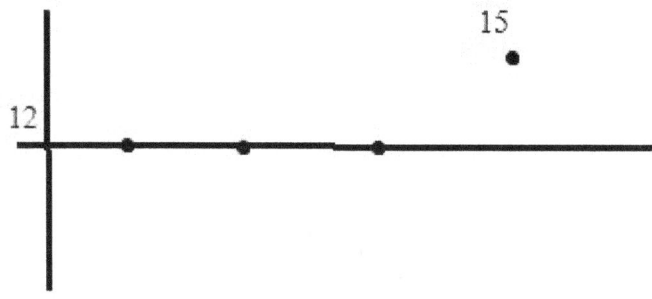

In the diagram above the fifth dot has been left off so that we can figure out where it would have to go. Because 15 is higher than the average, wouldn't the fifth dot have to be below the average, meaning below the line for 12? And because the fourth dot is exactly 3 units too high, wouldn't the fifth dot have to be 3 units below 12, which would put it at 9?

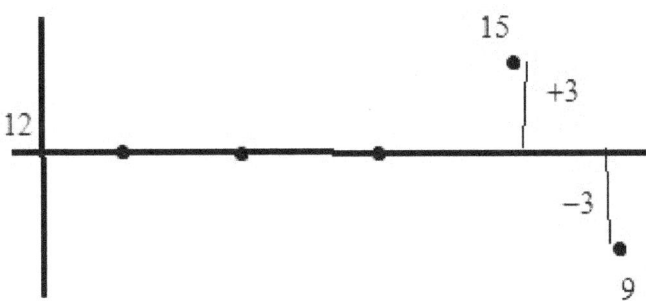

This is a wonderful tool for organizing averages! You may not need it for this particular question but it's a great way to start thinking about averages.

Getting back to this question, there are two ways that you can approach it. You could try to figure out what set of four numbers has an average of 8 and a median of 7 or you could simply test each of the answer choices. It wouldn't be too hard to make sure each answer choice did average 8 and it wouldn't be too hard to find the median for each set.

Many test questions require you to decide whether to find the solution first or test the answer choices first. Sometimes it's hard to know which will be fastest. If one approach fails, you can try the other. For this question, because it looks like testing the answer choices won't be too hard, let's start with that.

Choice A. First we have to test that the average is 8. Consider that if four numbers have an average of 8, they act as though they all were 8. That means they have to add up to 4*8 or 32. To add the numbers in choice A, let's add them in an order that makes the addition easy. First add 9 and 11, to get 20. Adding 3 and 5 to that is only 28. Choice A is out because the numbers don't average 8.

Choice B. Add the numbers in choice B by adding 8 and 12 first. That's 20. 6 plus 6 is 12. 20 + 12 = 32, so these numbers do average 8. Now find the median.

	6
	6
Median->	
	8
	12

Organizing the numbers visually shows us that the median is halfway between 6 and 8, making it 7. Choice B satisfies the conditions of the question and must be the answer.

Just as an experiment, let's see what would happen if you tried to create a series that met the conditions of the question, instead of testing the answer choices. We need a good visual tool for doing that. Let's try using the table above.

Chapter 4. ACT®-style Questions with Explanations

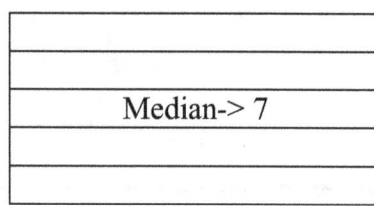

We know that there will be four numbers and that the median must be half way between the second and third number. The median is 7, so the second and third numbers could be one more and less than 7, two more or less than 7, and so on. As you can see, there are an infinite number of possibilities. Let's try making the second and third numbers one more and less than 7.

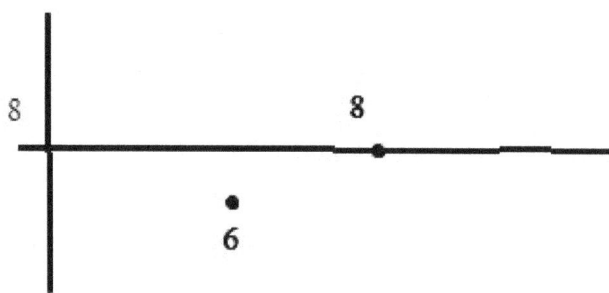

You may notice that because of how we have done this so far, the average of the three numbers is 7. To get to an average of 8 we need to have bigger numbers. However, the first number has to be less than or equal to 6, so the fourth number will have to be much larger.

At this point it might become clear that there are too many ways to create this combination and that it's not likely that any of them would actually be in the answer choices. So it turns out that testing the answers was the better way to go.

As an exercise to build your skills, though, you can try plugging in some numbers that would work in the above example. Use the visual aid that we used earlier on, in order to find an average of 8. Remember that the first number must be either lower than 6 or equal to 6.

8. A square shed sits on the boundary of a square lot. The floor of the shed has an area of 72 square meters and the lot has an area of 162 square meters.

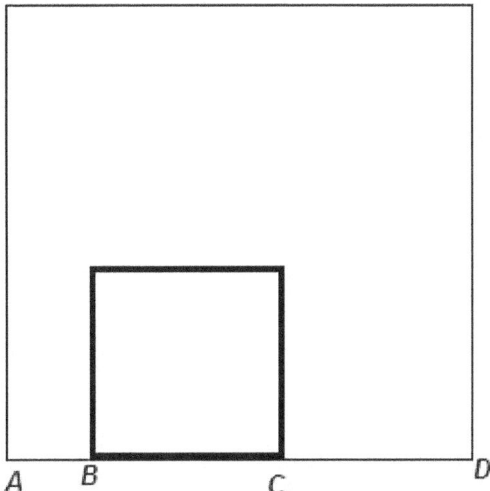

If points A and D are corners of the lot and points B and C are corners of the shed, what is the combined distance in meters of $\overline{AB}$ and $\overline{CD}$?

F. $2\sqrt{2}$
G. $3\sqrt{2}$
H. $6\sqrt{3}$
J. 45
K. 90

Explanation of 8:

Orient yourself carefully to what the question is asking. You need to find the combined length of two segments that are separated by another segment. Do you see the trick here?

There is no way to determine the length of $\overline{AB}$ or of $\overline{CD}$ individually. You can prove this to yourself by imagining moving the shed closer to point A or further from point A. The distances of $\overline{AB}$ and $\overline{CD}$ will change. Because the figure cannot be assumed to be drawn to scale, we simply can't know exactly where on $\overline{AD}$ the shed is located.

To get the answer, find the length of $\overline{AD}$ and then subtract from that the length of $\overline{BC}$. Those lengths **are** determined by the information in the question.

The problem requires a series of steps, so let's create a road map so as not to lose sight of the big picture.

Step 1. Calculate $\overline{AD}$. It is the square root of 162 (area of lot).
Step 2. Calculate $\overline{BC}$. It is the square root of 72 (area of floor of shed).
Step 3. Subtract step 2 and from step 1.

A road map is something that you actually write out on your scratch paper. It is important that it is visual and that it makes sense to you. It may seem a little simplistic or unnecessary but many errors on the test are due to the test taker getting lost in the details and forgetting what it is that they need to do. Lots of people working on this question might forget to add the two numbers together, or they might forget to take the square root of the areas. In the middle of the test your brain is easily confused. Intuitive tools like writing out a road map are critical for building your accuracy.

You don't need a road map for every question. If a question requires a number of steps, a road map may save you from an error.

Step 1. Calculate the square root of 162, which is the area of the square lot, in order to find the length of one side of the lot. Often, the best way to get a square root is to break the number into its prime factors.

Here's a good process for breaking numbers into prime factors. Write the number down.

162				

Start with the lowest prime number after 1 that you have not yet tested and ask yourself if that number goes into 162. We will start with 2. Does 2 go into 162? Yes. 2 into 162 is 81.

2				
162	81			

The top row shows the prime factors. The bottom row shows the remainder that you now break into prime factors. We start the whole process again with 81. Does 2 go into 81? No. Does 3 go into 81? Yes, 27 times.

2	3			
162	81	27		

Continue the process until the bottom row is a prime number.

2	3	3	3	3
162	81	27	9	3

If it helps, you can cross out each number in the bottom row as you break it down.

2				
~~162~~	81			

The numbers in the top row are the prime factors. The numbers in the bottom row are remainders, which you then try to break into prime factors. Experiment with a method of doing this that works best for you.

Using the above method, we have determined that the prime factors of 162 are:

$$2 * 3 * 3 * 3 * 3$$

Review your road map for step 1. We are trying to find the square root of this number.

$$\sqrt{2 * 3 * 3 * 3 * 3}$$

Let's regroup the numbers under the square root.

$$\sqrt{(3*3)(3*3)*2}$$

This can be rewritten as:

$$\sqrt{(3*3)} * \sqrt{(3*3)} * \sqrt{2}$$

Do you see that $\sqrt{(3*3)}$ is 3? You can test it by noticing that the square root of 9 is indeed 3. However, there is a much more important concept going on here. Let's take a closer look. (Fortunately, we have our road map to come back to after this digression.)

A powerful concept for working with square roots is to consider square roots to be a sort of parallel process to dividing something in half. If you want to divide a pizza with 8 slices in half, you need to end up with two piles, each with the same number of slices, namely 4 and 4. If you add the two piles together, you get the original number of 8 slices.

Similarly, taking a square root is a process of ending up with two piles that are exactly equal. However, instead of adding the two piles together to make sure you still have the same number, you multiply the piles. In other words, the square root consists of two exactly equal numbers that, when multiplied, give the original number.

For example, half of 16 is 8. 8 + 8 = 16. The square root of 16 is 4. 4 * 4 = 16.

Because of this, to find the square root of a number I just have to represent the number as the product of two other numbers with the same value.

16 = 4 * 4, so the square root of 16 is 4.

Similarly, you can write $\sqrt{4*4}$ is 4. Instead of writing 16 under the square root sign, I broke 16 down into its two equal parts, which instantly tells me what the value of the square root is.

Coming back to step 1 of our problem, we now have determined that the square root of 162 can be written as:

$$\sqrt{(3*3)} * \sqrt{(3*3)} * \sqrt{2}$$

Notice that we rewrote this purposely to group pairs of numbers together. This automatically shows us square roots.

$$3 * 3 * \sqrt{2} = 9\sqrt{2}$$

Step 1 is now complete. For step 2 we need the square root of 72. Find the prime factors of 72.

2	2	2	3	3
72	36	18	9	

The prime factors are 2 * 2 * 2 * 3 * 3. The square root of this is:

$$\sqrt{(2*2)(3*3)*2}$$

$$= \sqrt{(2*2)} * \sqrt{(3*3)} * \sqrt{2}$$

$$= 2 * 3 * \sqrt{2}$$
$$= 6\sqrt{2}$$

Step 3 of the roadmap is to subtract step 2 from step 1.

$$9\sqrt{2} - 6\sqrt{2} = 3\sqrt{2}$$

Choice G is the correct answer.

---

9. Justin decides to measure his new locker, which is in the shape of a rectangular prism. It measures 18 inches across, 24 inches deep, and has a volume of 13 cubic feet. How high is the locker in inches?

(Note: One cubic foot is equivalent to 1,728 cubic inches.)

A. 35
B. 45
C. 52
D. 84
E. 173

Explanation of 9:

Orient to this question carefully. It is more common for a math question to give you three dimensions and ask you to calculate the volume. Here they are giving you two dimensions and the volume. You are being asked to calculate the third dimension.

Also notice that the question switches units. It contains both inches and feet. It is easy to get confused about this. You may want to make a note on your scratch paper to watch out for the difference.

The relationship between width, depth, height and volume is:

$$w * d * h = v$$

or, in this case

$$18 \text{ inches} * 24 \text{ inches} * h \text{ inches} = 13 \text{ cu ft}$$

Your job is to solve for *h*, which is supposed to be in inches. You have two options. You can convert cubic feet into cubic inches to obtain an answer in inches or you can convert inches into feet to obtain an answer in feet, which you then have to convert back into inches.

Both 18 and 24 inches can convert fairly easily into feet so that may be easier than multiplying 13 cubic feet times 1,728 cubic inches per cubic foot.

$$18 \text{ inches} = 1.5 \text{ feet}, 24 \text{ inches} = 2 \text{ feet}$$

$$1.5 * 2 * h = 13$$

$$3h = 13$$

$$h = \frac{13 \text{ feet}}{3} * \frac{12 \text{ inches}}{1 \text{ foot}} = \frac{13 * 12}{3} \text{ inches}$$

Even though this calculation looks complicated, it can be simplified by regrouping.

$$13 * \frac{12}{3} = 13 * 4 = 13 * 2 * 2 = 26 * 2 = 52$$

Notice how we have regrouped the numbers so that the calculations are simple and intuitive. Choice C is the correct answer.

What would have happened if you had chosen to convert the cubic feet to inches? The resulting equation would be

$$18 \text{ inches} * 24 \text{ inches} * h \text{ inches} = 13 \text{ cu ft} * \frac{1728 cubic\_inches}{cubic\_foot}$$

or

$$h \text{ inches} = \frac{13 * 1728}{18 * 24}$$

Rather than use your calculator to generate huge numbers, which can lead to errors, break 1728 into prime factors. There is a good chance that prime factors in the denominator will cancel out some of the prime factors in the numerator and the calculation will become much simpler than it appears. Use the method below to break 1728 into prime factors.

2	2	2	2	2	2	3	3	3
1728	864	432	216	108	54	27	9	

Below is the equation will all numbers represented by their prime factors.

$$\frac{13 * 2 * 2 * 2 * 2 * 2 * 2 * 3 * 3 * 3}{(2 * 3 * 3) * (2 * 2 * 2 * 3)}$$

There are three 3's on the top and 3 on the bottom, so all the 3's cancel out. There are four 2's on the bottom and six on the top, so all of the bottom 2's cancel out and there are two 2's left on the top.

$$\frac{13 * 2 * 2}{1} = \frac{26 * 2}{1} = 52$$

Notice how we've regrouped the numbers above to do easy multiplications (13*2) rather than hard ones (13*4).

Using prime factors to simplify the division is very direct, intuitive and fool proof!

---

10. Annie studied the probability of a certain combination of cards occurring in a standard game of poker. She found that the probability depended on the number of players and could be determined by the equation $\frac{5^n k}{n!}$, where k is a constant and *n* is the number of players. If the probability of the combination of cards with 2 players is 0.5, approximately what is the probability of the combination with 3 players?

F. 0.83
G. 0.75
H. 0.33
J. 0.25
K. 0.16

Explanation of 10:

This question is easier than it looks. Remember that the test often tries to intimidate you. Don't get fooled by appearances. Let's take a closer look at the question.

First, you have to understand the symbol in the denominator. The exclamation point (!) stands for "factorial", which is more easily explained by examples than by words.

$$5! = 5*4*3*2*1$$

$$10! = 10*9*8*7*6*5*4*3*2*1$$

$$3! = 3*2*1$$

The one, of course, is not strictly necessary but it helps you see the pattern.

The problem gives you the value of the equation for n=2. You can use that to determine the value of k.

$$\frac{5^2}{2*1}k = \frac{25k}{2} = 0.5$$

$$25k = 1$$

$$k = \frac{1}{25}$$

The question asks you to determine the probability for n = 3.

$$\frac{5^3}{3*2*1} * \frac{1}{25} = \frac{5*5*5}{3*2*5*5} = \frac{5}{6}$$

Notice how we have used prime factors to simplify this calculation. The two 5's in the denominator cancel out two 5's in the numerator.

We now have an answer that is in fraction form but the answer choices are in decimal form. You could use a calculator or long division to find the decimal equivalent of $\frac{5}{6}$. However, it may be more intuitive and fool-proof to compare $\frac{5}{6}$ with the answer choices.

Notice that the answer choices are in order from largest to smallest. This is common on the test. Where does $\frac{5}{6}$ fall in that range? Use estimation. $\frac{5}{6}$ is more than half. Which answer choices are greater than

Chapter 4. ACT®-style Questions with Explanations

0.5? Only F and G. Choice F is difficult to work with. Choice G is exactly three fourths. Compare $\frac{5}{6}$ and $\frac{3}{4}$. First of all, they are not the same so it is unlikely that $\frac{3}{4}$ could be the right answer. If choice F is correct, then $\frac{5}{6}$ must be larger than $\frac{3}{4}$. Test it out. The common denominator is 24.

$$\frac{5}{6} * \frac{4}{4} = \frac{20}{24} \text{ versus } \frac{3}{4} * \frac{6}{6} = \frac{18}{24}$$

Choice F is correct.

You finished Day 2! How did it go on these five questions?

Number of questions you got right on your own: _____

Types of problems or patterns you need more work on: _____

_____

How much new did you learn from these questions? ☐ Important tools! ☐ Some tools ☐ Not too much

Day 3, Questions 11-15

11. Which of the following inequalities is equivalent to the expression $3x - 10 > 5x + 12$?

A. $x < 0$
B. $x > 11$
C. $x < -11$
D. $x > 2$
E. $x < -2$

Explanation of 11:

This question is pretty straightforward. You need to rewrite the original inequality in a form that matches one of the answers. Look at the answer choices. They are all expressed in terms of x being greater than or less than a number. This tells you how you need to manipulate the original inequality.

It is important to remember that you can apply all of the manipulations for equations to inequalities with one critical exception. You can add or subtract the same amount from each side without changing the validity of the inequality, just as you can with an equation. You can multiply or divide each side by the same nonzero number but **only** if the number is positive.

It is possible to multiply or divide both sides of an inequality by a negative number but if you do so, you must reverse the inequality. For this reason, you can't multiply or divide both sides of an inequality by a variable unless you know whether the variable is positive or negative.

Most questions that involve inequalities will test you on this. In this question you have to use a little algebra to get all the x's on one side and all of the numbers on the other. Your work should look something like this:

$3x - 10 > 5x + 12$	
$-3x + 3x - 10 > 5x + 12 + -3x$	adding $-3x$ to both sides
$-10 > 2x + 12$	simplifying
$-12 + -10 > 2x + 12 + -12$	adding -12 to both sides
$-22 > 2x$	simplifying
$-11 > x$	dividing both sides by 2
$x < -11$	rewriting so that x is on the left

Notice that in the second to the last step we divided both sides by 2. Because 2 is positive, the inequality sign remains the same. In the final step there is no algebraic manipulation. We have simply written the inequality with the x on the left. It does take a bit of care to make sure that you do not get the direction of the inequality confused when you do this.

Choice C is correct.

It is possible to address this problem by testing answer choices. Assign a value for x that matches the answer choice. If you do that, you will find that choices B and D are out because any positive number greater than 2 or 11 will create a false statement in the original inequality. However, when you test choice A, let's say $x = -20$, it will work. If you choose $x = -5$, it will not. The lesson is that if you test answer choices, more than one may appear to work and you have to do something else to eliminate the incorrect ones.

---

12. Which of the following expressions is equivalent to $x(5 - x) - 3(x - 11)$?

F. $-x^2 + 2x + 33$
G. $-4x^3 + 2x + 33$
H. $x^2 - 2x - 33$
J. $x^2 + 2x - 33$
K. $-x^2 + 33$

Explanation of 12:

As you orient yourself to the question stem, consider what is going to make it difficult for you to be accurate. For most people the subtractions will increase your chances of making a mistake.

There is a powerful strategy for avoiding such mistakes! Replace subtractions with additions. Consider that -5 is the same as $-1(5)$. So the equation $x - 5$ can be rewritten as $x + -1*5$, or $x + -5$. Instead of subtracting, you are adding negative numbers.

Chapter 4. ACT®-style Questions with Explanations

Let's rewrite the original expression replacing subtractions with additions.

$$x(5 - x) - 3(x - 11)$$
$$x(5 + -1x) + -3(x + -11)$$

Now expand the expression.

$$x5 + x*-1*x + -3x + -3*-11$$

Notice that we are doing this one step at a time. Avoid the tendency to do multiple steps in your head. Doing one step at a time and doing it on paper is a key way to increase your accuracy.

$$x5 + x*-1*x + -3x + -3*-11$$
$$5x + -1x^2 + -3x + 33$$

In the second line above, note that –3 times –11 yields a positive number. Continue simplifying.

$$-x^2 + 2x + 33$$

This matches choice A, which is the correct answer.

There is another approach that you should be aware of for working with questions that have variables in the answer choices. It involves simply assigning a value to x. In this case, suppose we set x equal to 2. Then, instead of variables, we have numbers, which are easier to work with.

$$2(5-2) - 3(2-11)$$
$$2*3 - 3*-9$$
$$6 - -27$$
$$6 + 27$$
$$33$$

The plan here is that when x is 2, the answer comes out to 33. You can now test each answer choice by setting x equal to 2.

There are two drawbacks to this method. The first is that you now have to test all answer choices. For some problems this is still faster and more accurate than other methods. For this problem, it is not.

The second drawback is that more than one answer choice might work. There is only one answer that works for **every** value of x but there may be more than one that works for the value of x that you chose. For that reason it is best to avoid certain values of x, such as 0 or 1, that have special properties.

If more than one answer choice works, those that did not work are definitely out but in order to find the one correct answer, you have to choose a different value for x.

In this problem it was easy enough to just manipulate the original expression by eliminating subtraction, so it was not necessary to use assigning a value to the variable.

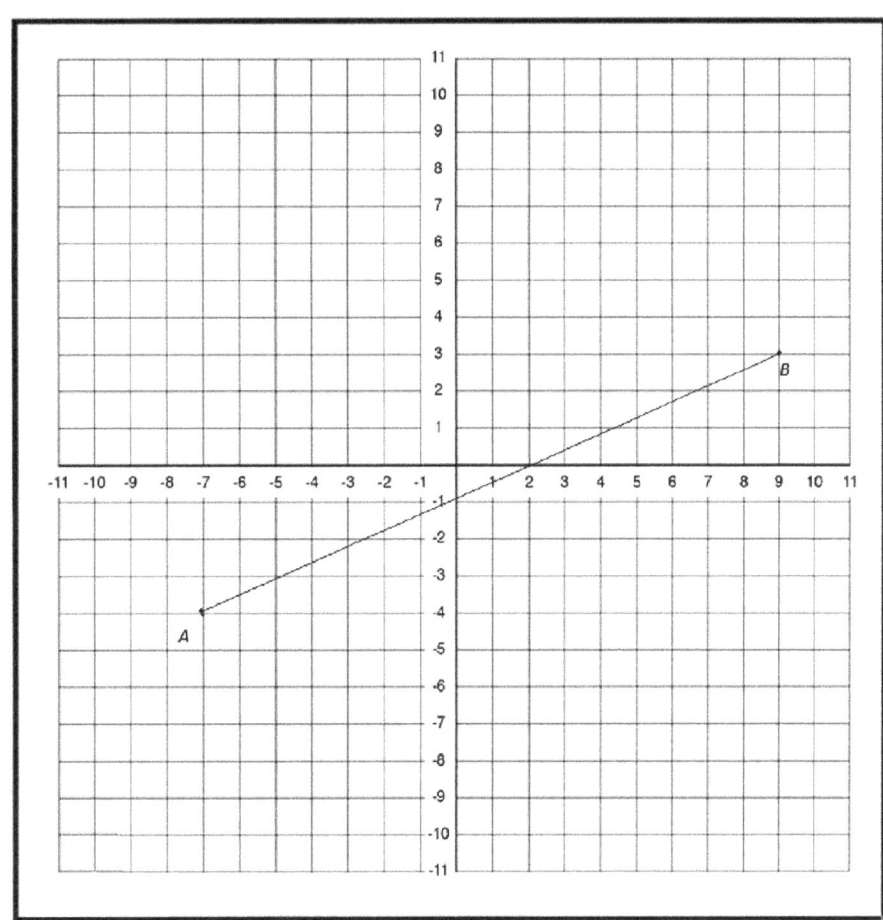

13. In the graph above, the coordinates of point A are (-7,-4) and the coordinates of point B are (9,3). What is the *x*-coordinate of the midpoint of line segment $\overline{AB}$?

A. –1
B. 0
C. 1
D. 3.5
E. 8.5

Explanation of 13:

There is a hard way to do this problem and an easy way. The hard way is to figure out how long $\overline{AB}$ is, divide that number in half and then go that many units up from point A along the segment (or down from point B) and calculate the coordinates of the resulting point.

Chapter 4. ACT®-style Questions with Explanations

The easy way is easier but it is a little complicated to explain why it works. Imagine that $\overline{AB}$ is the hypotenuse of a triangle with a base starting at point A and parallel to the *x*-axis. The other end of the base would be directly below point B. Let's call that point C. The mid point of the hypotenuse is directly above the midpoint of the base. If you find the *x* coordinate of the midpoint of the base, that will also be the *x* coordinate of the midpoint of $\overline{AB}$.

The coordinates of point C are (9,-4). The length of $\overline{AC}$ is the distance from point A (-7,-4) to point C. From –7 to 9 is 16 units. Half of that is 8 units. 8 units from point A is (1,-4). This is also 8 units from C. The *x* coordinate is 1 and so the *x* coordinate of the midpoint of $\overline{AB}$ is also 1. Choice C is the correct answer.

Do you see why? If you draw a line from the midpoint of $\overline{AC}$ straight up to the hypotenuse, you are really drawing a line along the line $x = 1$.

Another way to understand why the midpoint of the base is directly below the midpoint of the hypotenuse is to consider the concept of rise over run. The rise of the line over the run of the line is the slope, which is constant for the entire line.

The run from point A to point C is 16 units and results in a certain rise. If you run half of that distance, you will rise half that distance. That means you will be at the midpoint of the base and you will also have risen to the midpoint of the line $\overline{AB}$.

---

14. Which of the following expressions is equivalent to $(5x^2 + 3y^3)(5x^2 - 3y^3)$?

F. $25x^4 + 16x^2y^2 - 9y^6$
G. $25x^4 - 16x^2y^2 - 9y^3$
H. $25x^4 - 9y^6$
J. $10x^4 - 9y^6$
K. $10x^4 - 16x^2y^2 + 9y^6$

Explanation of 14:

There are two ways to approach this. One is to assign values to *x* and *y* and then test answer choices. However, because the answer choices involve such large exponents, this would probably require too much calculation.

The second method is to try to simplify the original expression. Take note of what is going to be confusing for you so you can be especially careful. Most people will find multiplying exponents to be confusing.

The antidote for confusion is to make calculations step by step and write them down. On paper! Not in our heads. We can start by FOILing the original expression.

$$(5x^2 + 3y^3)(5x^2 - 3y^3)$$

$$(5x^2 * 5x^2) + (5x^2 * -3y^3) + (3y^3 * 5x^2) + (3y^3 * -3y^3)$$

$$(5*5* x^2 * x^2) + (5 * -3 * x^2 * y^3) + (3 * 5 * x^2 * y^3) + (3 * -3 * y^3 * y^3)$$

At this point there may be some confusion as to how to multiply these variables with exponents. For example, does $(y^3 * y^3)$ come out in terms of y^3 or y^6 or y^9? There is an easy way to remind yourself. Write out y^3 as $y*y*y$.

$$(y^3 * y^3) = y*y*y * y*y*y$$

Is it clear now that this is y^6? You can think of y^3 as a string of y's connected together by multiplication, just like a string of beads. If you have a string of three and connect it (with multiplication) to another string of three, you have a string of six.

Now it should be easy to simplify the original expression.

$$(5*5* x^2 * x^2) + (5 * -3 * x^2 * y^3) + (3 * 5 * x^2 * y^3) + (3 * -3 * y^3 * y^3)$$
$$25x^4 + -15 x^2 y^3 + 15 x^2 y^3 + -9 y^6$$

Notice that the middle terms cancel out. What remains is the same as choice H, the correct answer.

---

15. The line shown below in the standard (x,y) coordinate plane passes through the points $(-3,4)$ and $(3,1)$. What is the slope of the line?

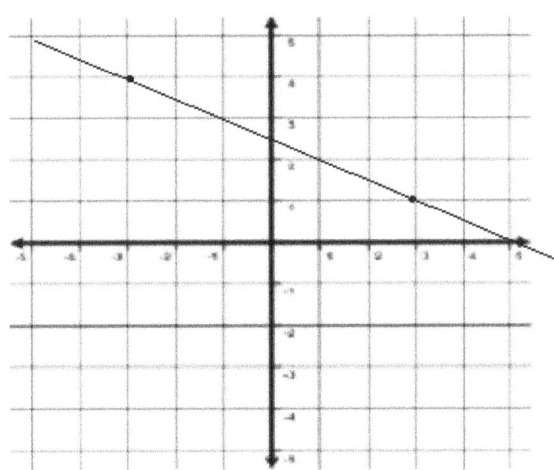

A. $-\dfrac{9}{4}$

B. -1

C. $-\dfrac{1}{2}$

D. $\dfrac{1}{2}$

E. $\dfrac{9}{4}$

Explanation of 15:

This is a relatively simple slope problem but it is a good opportunity to review some important aspects of slope.

Slope is the rise of a line over the run. The run means the amount of distance along the *x*-axis and it must be measured going from left to right. In the diagram for this problem the run from the leftmost point to the rightmost point is 6 units. The run is always a positive number, as long as you go from left to right.

The rise measures how much the line has dropped or risen – along the *y*-axis – as you go along the run. Going left to right, if the line has dropped, the rise is negative.

In our case the line has dropped from $y = 4$ to $y = 1$, a rise of -3.

The rise over the run is $\dfrac{-3}{6}$ or $-\dfrac{1}{2}$. Choice C is correct.

You finished Day 3! How did it go on these five questions?

Number of questions you got right on your own: _____

Types of problems or patterns you need more work on: _____

_____

How much new did you learn from these questions? ☐ Important tools! ☐ Some tools ☐ Not too much

Day 4, Questions 16-20

16. Four points are plotted on a number line. Point M is at –3.

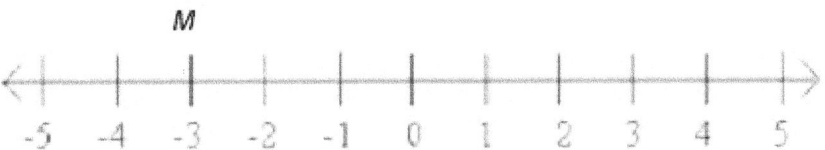

Point N is greater than point M but less than point O. Point P is less than point N but greater than point M. Which of the following relationships *must* be true of the lengths of the line segments?

F. $\overline{MO} > 3$
G. $\overline{MP} < 0$
H. $\overline{MP} < \overline{NO}$
J. $\overline{PO} > \overline{MN}$
K. $\overline{PO} > \overline{NO}$

Explanation of 16:

Orient to the question stem. The stem asks which answer choice *must* be true. The word "must" is italicized because it is critical. If the correct answer must be true, what does that mean about the incorrect answers?

If you said, "They must be false," you have a lot of company but that is not correct. The wrong answer **could** be false or they might be true but they do not **have to** be true. This is an important distinction.

Start out by placing points N, O, and P on the number line. The trick here is that you don't know exactly where the points are. You only know their relative positions but in order to put them on the number line, you have to arbitrarily place them at a specific position.

Start by putting N somewhere to the right of M. Leave enough room to put other points if necessary. Point O is even further to the right than N. Point P is to the left of N but to the right of M.

Chapter 4. ACT®-style Questions with Explanations

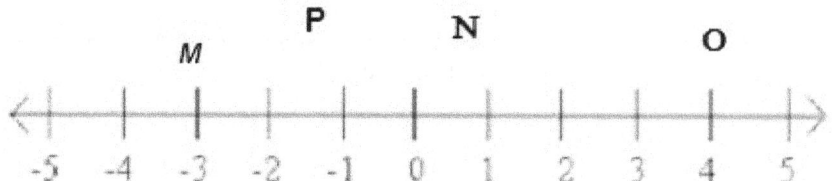

The challenge in this problem is to be aware of what you do **not** know. You don't know how far it is from any one point to another. That means you don't know if P is very close to M or P is far from M. You only know the relative positions.

Your strategy now is to test the answer choices. For choice F, must $\overline{MO}$ be greater than 3? Test this by showing that O might be only a unit or two from M. Points P and N can still fit into their relative positions, so there is no reason that $\overline{MO}$ must be greater than 3.

For choice G, $\overline{MP}$ being less than zero would mean that P is to the left of M, which is not true. For choice H, $\overline{MP}$ could be less than $\overline{NO}$ but it does not **have to** be. You could redraw the diagram with a longer distance between M and P and a shorter distance between N and O.

Choice J also could be true but does not have to be. You could redraw the image to make M and N very far apart and make P and O closer.

For choice K, can you see from the diagram that $\overline{NO}$ is a part of the segment $\overline{PO}$? The part must be shorter than the whole segment. Choice K is the correct answer.

---

17. On the standard (x,y) coordinate plane, for the graph of the function $f(x) = \sin^2 x + \cos^2 x$, for values of $x \leq \frac{\pi}{2}$ through $x \geq -\frac{\pi}{2}$, which of the following represents the midpoint of the graph?

A. (0,0)
B. $(\frac{\pi}{4}, 1)$
C. $(1, \frac{\pi}{4})$
D. (0,1)
E. (1,0)

Explanation of 17:

This question is not anywhere near as complicated as it looks. Don't despair! There are two simple facts that you need to learn and these will take you to the answer.

First, $\sin^2 x + \cos^2 x$ is a special relationship. It equals 1. This means that the f(x) = 1, which is the same as y = 1. The graph of that is simply a line parallel to the x-axis and one unit above the x-axis.

Before considering the rest of the problem, let's look at why $\sin^2 x + \cos^2 x$ equals 1. It's not as mysterious as it may seem. In working with tangents, sines, and cosines, it's often helpful to go back to their definitions.

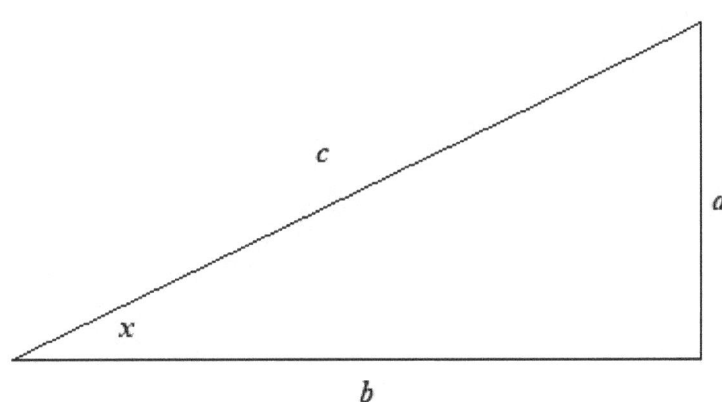

For the angle x above, the sine is opposite over hypotenuse, or $\frac{a}{c}$. The cosine is adjacent over hypotenuse, or $\frac{b}{c}$. Check out what happens when you square each of these and then add them. $\frac{a^2}{c^2} + \frac{b^2}{c^2} = \frac{a^2 + b^2}{c^2}$. The Pythagorean theorem tells us that $a^2 + b^2$ is equal to c^2, so the fraction is equal to 1. You can memorize $\sin^2 x + \cos^2 x$ equals 1 but now you know how to remind yourself if you forget.

The problem tells us that we are looking at only a segment of the line, for values of x between and including $\frac{\pi}{2}$ and $-\frac{\pi}{2}$. Where are these on the x-axis? In reality it doesn't matter exactly where they are.

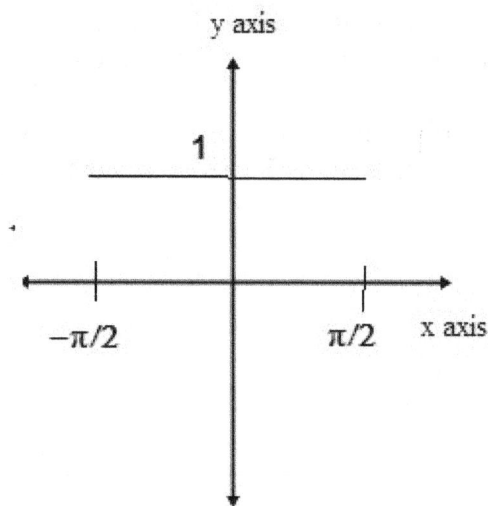

Notice in the image above that $\frac{\pi}{2}$ and $-\frac{\pi}{2}$ are equidistant from the y-axis. You don't need to know how far they are from the y-axis. You only need to know that the half way point between them is $x = 0$, which is the y-axis.

Consider the two facts that you now know about the midpoint of the graph $f(x) = \sin^2 x + \cos^2 x$. Its x coordinate is 0. Its y coordinate is 1. The midpoint is $(0,1)$, which is choice D.

There are some clues that can make this problem even easier. Once you know that the y coordinate is 1, there are only two possible answer choices, B and D. This gives you a 50:50 chance.

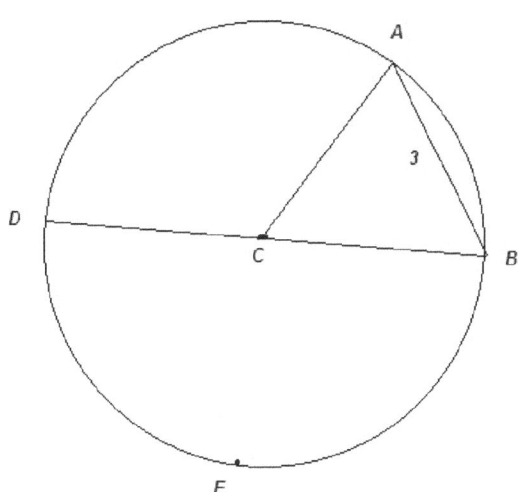

18. In the diagram above $\overline{DB}$ is a diameter of the circle with center C. The triangle ACB is isosceles and has a perimeter of 10. Which of the following is the length of arc DEB?

F. 8π
G. 7π
H. 4π
J. 3.5π
K. 3π

Explanation of 18:

This problem has several elements that you need to be well oriented to. Using the KING approach, first make a note of what you know. $\overline{DB}$ is a diameter. Triangle ACB is isosceles. $\overline{AB}$ is 3. The perimeter is 10. Next consider what else you can infer from what you know. Arc DEB is actually half the circumference. Third, consider what you need to know to get to the answer. To find the circumference, you need to know the radius or diameter. To find the radius, you need to know the length of $\overline{CB}$. Fourth, consider how you get the information you need. To find $\overline{CB}$, you need to use the facts that the triangle is isosceles and that one side is 3.

This involves a number of steps so this is a good place to use a road map.

    Step 1. Determine the length of $\overline{CB}$.
    Step 2. $\overline{CB}$ is a radius. Calculate the circumference of the circle using $2\pi r$.
    Step 3. The answer is half of step 2.

Given that triangle ACB is isosceles, how can you figure out the length of $\overline{CB}$? In an isosceles triangle two sides are the same but is $\overline{AB}$ (which equals 3) one of the sides that is the same or is it the side that's "left out."?

Remember the isosceles triangle trap. If one side of an isosceles triangle is 3 and the perimeter is 10, do you see that there are two options?

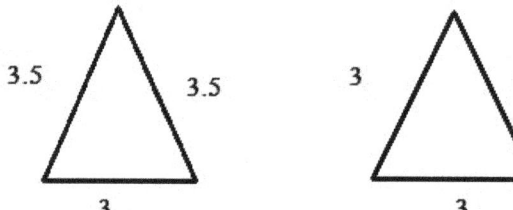

For the triangle on the left, we assume that the 3 is the side that is "left out." Often when an isosceles triangle is drawn, the base is the odd side and the two sides are identical. The triangle on the left is drawn that way. However, this can be a trap! The test likes to make you think that the base is the side that's out but an isosceles triangle can also be drawn as on the right. The base is one of the matching sides.

In fact, when you see an isosceles triangle on the test, more often than not they are trying to make you fall into this trap. Remember that you do NOT know which side is the paired side and consider both options.

If there are two ways that triangle ACB could be drawn, how can you figure out $\overline{CB}$? There is another clue! Look carefully at the original diagram. Sides $\overline{AC}$ and $\overline{CB}$ are both radii of the circle. They must be the same. The side of 3 is the odd side out. The perimeter of 10 minus the side of 3 leaves 7, which must be divided equally between the two other sides. $\overline{CB}$ must be 3.5.

Now go back to your road map (which, if you are like most people, you have probably forgotten by now.) Step 1 is 3.5. Step 2 gives us 7π. Step 3 gives us 3.5π. Choice J is the correct answer.

---

19. At the beginning of a school year, the administration of a certain high school required all registered students to respond to the question "Would you rather have beef or chicken dishes at lunch?" The results showed that most students preferred chicken. A student organization questioned the results and asked the first 100 students that arrived on a Friday to respond to the question "Would you prefer red meat, fish, poultry, or vegetarian lunches?" The majority of the students preferred vegetarian meals. As the school year went on, the cafeteria found that most students chose vegetarian meals. Which of the following best describes this situation?

A. A randomized survey can be more accurate than a census.
B. A census can be more accurate than a nonrandomized survey.
C. An experiment can be more accurate than a randomized census.
D. A randomized survey can be more accurate than a randomized census.
E. A nonrandomized experiment can be more accurate than a randomized experiment.

Explanation of 19:

This is a bit of an unusual question type but it has appeared on tests. Examine the answer choices to see what things you need to distinguish. Some questions refer to surveys, others to censuses, and others to experiments. To solve the problem you must consider how these differ.

Secondly, the questions ask about randomized and nonrandomized events. Consider what this means.

Let's deal with the second distinction first. Does the passage talk about a randomized or a nonrandomized situation? Actually, the passage describes two different situations – the one conducted by the administration at the beginning of the year and the one conducted by the student organization. The administration's questionnaire was given to all students. That is not random. It includes everyone. The student organization questioned the first hundred students who showed up. The students were not chosen based on any criteria related to the topic, so it can be considered a random selection.

Let's now consider the first distinction. Was the administration's questionnaire a survey, a census, or an experiment? An experiment tests an hypothesis by doing something and then measuring the result. The administration did not do this. They asked a question. Is that a survey or a census? A census is defined as a gathering of data about an entire population. The administration's question does constitute a census because they questioned everyone.

The student organization conducted a survey because they did not question the entire population. The two events, then, are a nonrandomized census and a randomized survey. The passage implies that the student organization's survey was more accurate.

	Type	**Random**	**Accuracy**
Administration	Census	Nonrandom	No
Student organization	Survey	Random	Yes

Choices C and E are out because they refer to experiments. Neither of the above is an experiment. Choice A involves a randomized survey (student organization) and a census (administration.) It says that the former was better. This matches our data. Let's check the remaining answer choices just to be sure.

Choice B refers to a nonrandomized survey, but the survey (student organization) was random. Choice D refers to a randomized survey (student organization) but compares it to a randomized census. The census was not randomized. By definition a census is not randomized. It tries to count everyone. Choice A is confirmed as the correct answer.

If you run across a question like this, the definitions that we have used for survey, census, and experiment should get you to the right answer.

20. Two runners, Amy and Jason, are running on a country road, as shown below.

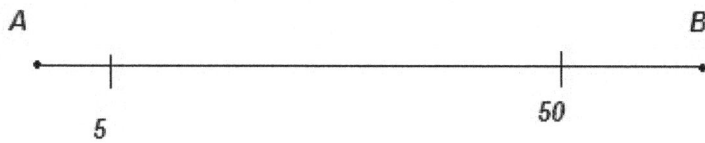

Amy starts at mile 5 and runs toward point B at 11 miles per hour. Jason starts at mile 50 and runs toward point A at 6 miles per hour. After how many hours, to the nearest tenth, will the two meet?

F. 7.5
G. 4.1
H. 3.7
J. 2.6
K. 2.0

Explanation of 20:

This type of problem is best solved with a "snapshot" approach. That means making a picture of what the situation will look like at certain points in time. What will it look like after 1 hour? What will it look like after 2 hours? This step-by-step approach helps keep you accurate and also helps you develop an intuitive sense of the situation.

What will this scenario look like after 1 hour? In hour Amy runs 11 miles, so she will end up at mile 16. Jason runs 6 miles in 1 hour. This takes him to mile marker 44.

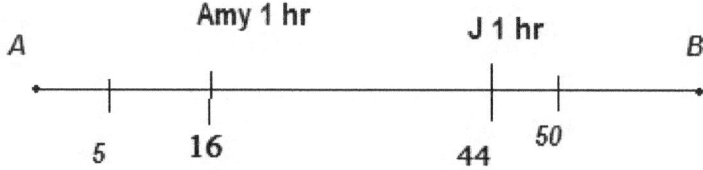

Clearly, they have not met after 1 hour. What will the picture look like after two hours? Amy runs 11 more miles to 27. Jason runs 6 more miles to 38. They still would not have met. However, you can see that they

78   Chapter 4. ACT®-style Questions with Explanations

are only 11 miles apart. In another hour Amy will have covered that distance and must have passed Jason. Three hours is too much. Looking at the answer choices, only choice J is bigger than 2 but less than 3.

If this is not intuitively obvious to you, you can draw out the third hour. However, first consider the answer choices. There is no choice for 3 hours. You know that choice K, 2 hours, is out. If the correct answer is 2.6 hours, then testing 3 hours will show that 3 is too much. If 2.6 is not the correct answer, then 3 hours will be too small.

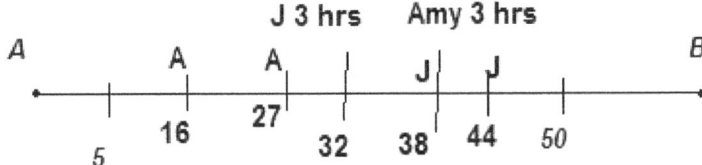

By using this format for showing each step, you can literally see where each person is and determine when they have met. Notice that you do not need to calculate what 2.6 hours looks like. You simply determine that 2 is too little and 3 is too much.

If you are a highly motivated math nerd, you can now try to determine the algebra that would get you the correct answer. But be sure to notice that you didn't need it!

You finished Day 4! How did it go on these five questions?

Number of questions you got right on your own: _____

Types of problems or patterns you need more work on: _____

_____

How much new did you learn from these questions? ☐ Important tools! ☐ Some tools ☐ Not too much

Day 5, Questions 21-25

21. A rectangular garden plot has a length of 13 feet and an area of 104 square feet. What is the perimeter of the garden in feet?

A. 7
B. 8
C. 16
D. 21
E. 42

Explanation of 21:

Start by drawing a picture. A picture is a powerful intuitive tool. It tells you much more than the words do because it allows you to see relationships.

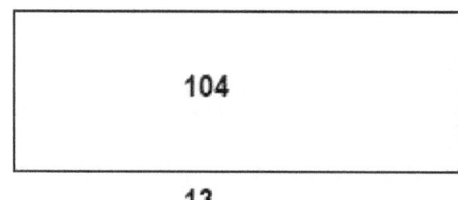

Using the KING approach, what do we Know? We know the length of one side and we know the area. What do we Need to know? The perimeter, which means we need to know the other side. How do we Get the other side? 13 times something equals 104.

Just to be safe, let's create a road map.

Step 1. Determine the length of the second side.
Step 2. Add together the lengths of all four sides – the 13 side twice and the second side twice.

Step 1. 104 divided by 13. You have a number of choices for doing this division. One is to use the calculator where allowed. If you are highly accurate with the calculator, this will be quick. Another method is to find the prime factors of 104 to see if it happens to be an exact multiple of 13.

You can use a table like the one below to find prime factors. Start by putting 104 in the bottom left space. Ask yourself if it is divisible by 2. It is and 2 goes into 104 52 times. Put 2 above the 104 and put 52 in the bottom row just to the right of 104. Repeat the same process for 52 and so on until you have found all of the prime factors.

2	2	2	13	
104	52	26	13	1

You now have $\dfrac{2*2*2*13}{13} = 2*2*2 = 8$. The second side is 8.

Step 2. The four sides of the rectangle are 13, 8, 13, and 8. You have lots of options for adding these together. One intuitive method is to group 8 and 13 together (21) and double it to get 42. You could also group the two 8's and the two 13's but it may be harder to add 16 and 26 than to double 21. When you look for an intuitive alternative to standard adding or multiplying, look for numbers that are easy to work with for you.

Our answer is 42 and choice E is correct.

22. Which of the following is a solution to the equation $x^2 - 64x + 4 = |-4|$?

F. -64
G. -16
H. 0
J. 16
K. 64

Explanation of 22:

Orient to the original equation. The question asks for a "solution". That means a value of x that makes the statement true. The equation involves a square so there are most likely two specific values for x that will work.

Look for the elements that make this equation a little tricky. Do you notice the absolute value on the right side? That definitely complicates things, so deal with that first. What is the absolute value of –4? Is there one value for that or are there two values? Be careful!

There is only one value that this can be. It must equal 4. Absolute values are always expressed as positive numbers. You can think of an absolute value as the distance that a number is on the number line from zero. 4 and –4 are both 4 units from zero. 4 is the absolute value for both of them.

$$4 = |4| \text{ and } |-4|$$

$|-4|$ can **only** be 4. It **cannot** be –4.

Having determined that $|-4|$ can only be 4, you can now rewrite the original equation as:

$$x^2 - 64x + 4 = 4$$

Before you get too involved with algebra now, let's use some good intuitive tools. Consider this:

$$\clubsuit + 4 = 4$$

What does ♣ have to be to make this work? Can you tell at a glance that it has to be 0?

$$0 + 4 = 4$$

It makes intuitive sense, right? This means that $x^2 - 64x$ must equal 0.

$$x^2 - 64x = 0$$

Chapter 4. ACT®-style Questions with Explanations

This process of looking at a part of the equation as a block or chunk is called chunking. You can chunk an equation by putting your thumb or finger over part of the equation so that, instead of seeing that part as a lot of details, you see it just as an intuitive chunk.

$$x^2 - 64x + 4 = 4$$

When you look at a "chunk", you are using a different part of your brain. Instead of seeing details, you are seeing the big picture. That uncovers certain obvious information that you couldn't see when you were focused on details.

Consider $x^2 - 64x = 0$. There are some ways to understand this intuitively.

$$\clubsuit - \blacklozenge = 0$$

What does this tell you about the two symbols? Do you see that they have to be equal? If I subtract one thing from another thing and there is nothing left over, the two things must be equal.

$$\clubsuit = \blacklozenge$$

This means that $x^2 = 64x$. You have some intuitive tools for understanding this expression (instead of getting hung up with algebra.) To see the intuitive connection you need to rewrite the x^2 as $x*x$.

$$x * x = 64 * x$$

Do you see the intuitive insight?

$$\clubsuit * x = 64 * x$$

If the left side is to look like the right side, doesn't $\clubsuit$ have to be 64? If $x * x = 64 * x$, then x must be 64.

$$64 * 64 = 64 * 64$$

If the intuitive strategies don't quite do it for you, let's take a look at more standard algebraic ways to work with $x^2 - 64x = 0$.

By adding $64x$ to both sides, you get: $x^2 = 64x$
By dividing both sides by x, you get: $x = 64$

Choice K is the correct answer.

23. Line segment $\overline{MN}$ is graphed on the standard (x,y) coordinate plane. Point M is (2,5). The midpoint of $\overline{MN}$ is (-3,-2). What are the coordinates of point N?

A. (-1,3)
B. (-4,-5)
C. (-8,-9)
D. (-6,-10)
E. (-9,-10)

Explanation of 23:

You will probably have a question like this that asks about the midpoint of a line segment graphed on the (x,y) coordinate plane. The key is to think in terms of slope, or more specifically rise over run.

Suppose you have a line segment with a slope of $\frac{1}{4}$. If you start at one end point of the line segment and "run" 4 units to the right and "rise" 1 unit up, you are at another point on the segment. The rise over run is 1 over 4.

If doing so takes you to the midpoint of the line segment, then to find the end of the line segment you simply go another 4 units to the right and another 1 unit up.

In this problem it will probably help to roughly graph the line segment so that you have good intuitive, visual information to keep you on track. Below is an example of a hand-drawn graph.

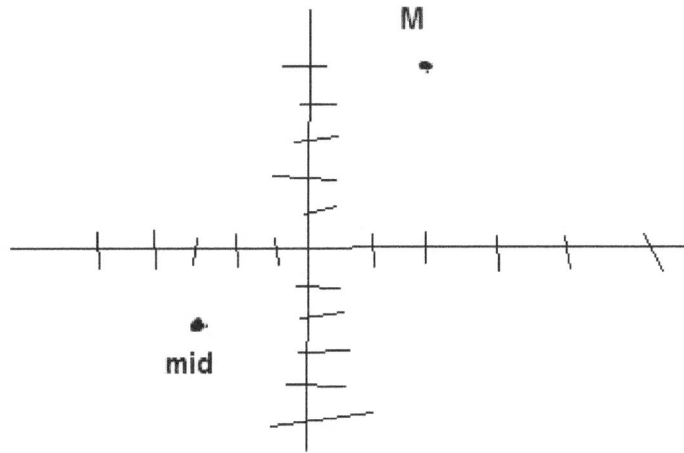

Now determine the rise and the run from point M to the midpoint. Because you are going from right to left, the run is negative. Because you are going down, the rise is negative. To get from M to the midpoint, you go 5 to the left and down 7. Notice that even with a rough drawing, you can count the number of units with your finger. If you go the same rise and run from the midpoint, you will arrive at the other end of the segment.

The midpoint is at (-3,-2). Going 5 to the left takes you to –8. Going down 7 takes you to –9. Choice C is correct.

Even though slope is usually stated as rise over run, it is better to calculate the run first and then the rise. This is because run is the *x* value and rise is the *y* value. It makes sense to figure out the *x* value first so you do not get confused between *x* and *y*.

---

24. The figure below shows a right triangle with lengths of the two sides shown in centimeters. Point D is the midpoint of the hypotenuse. What is the length of $\overline{DB}$ in centimeters?

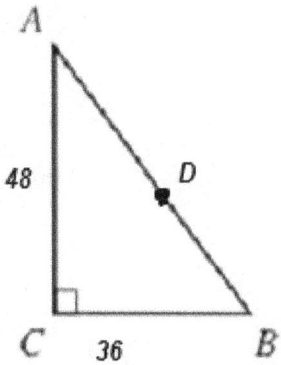

F. 18
G. 24
H. 30
J. 42
K. 60

Explanation of 24:

Orient to the question stem. It asks you to find the length of half the hypotenuse. You first half to find the length of the hypotenuse and then divide it in half. It would be easy to forget to divide the hypotenuse in half during the test. Your brain can easily get overloaded. If you notice the possible trap, avoid it by writing your road map. It may seem too simple to bother with right now, but in the middle of the exam people make such mistakes all the time!

    Step 1. Calculate length of hypotenuse
    Step 2. Divide step 1 in half.

You may remember the Pythagorean Theorem, which relates the lengths of the sides of the right triangle to the length of the hypotenuse - $a^2 + b^2 = c^2$, where *a* and *b* are the sides and *c* is the hypotenuse.

In this problem, though, the lengths of the sides are very large and will be cumbersome to square. You can do it with your calculator, where permitted, but the chances of making an error are large.

There is an important short cut that you can use on this problem. Do you remember 3:4:5 right triangles? If a right triangle has one side of 3 and one side of 4, the hypotenuse must be 5. The Pythagorean Theorem still applies to this but for these particular numbers, the square of 3 plus the square of 4 gives the square of 5 (9 + 16 = 25).

If you take a 3:4:5 triangle and blow it up to twice the size, keeping the proportions the same, each side will now be double. It will be a 6:8:10 triangle. The same holds true if you triple the size, and so on.

The test likes to give you 3:4:5 triangles in disguise. How can you determine whether the triangle in this problem is 3:4:5? Start with the shortest side, 36. It is 3 * 12. If this **is** a 3:4:5, then the next side would be 4 * 12, or 48. It is. The hypotenuse, then, must 5 * 12, or 60.

Here is a triangle. What is the length of side $\overline{AC}$?

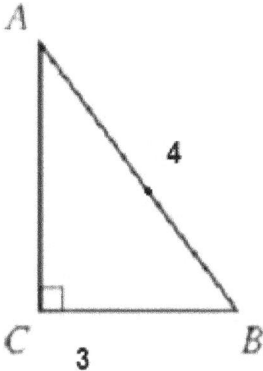

Watch out. This is a trap. Is it **not** 5? In a 3:4:5 right triangle, the 3 and 4 must be **sides**. In the above triangle, 4 is the length of the hypotenuse. Also remember that this only works in a right triangle.

Now we have the answer, 60, right? Sorry. No. Remember step 2. Half of 60 is 30. Choice H is correct. Half the people taking the test will forget to do the final step but you won't because you wrote out your roadmap!

25. If one cow is chosen randomly from a given herd, the odds that the cow has a black coat are 81:369. The odds of a randomly chosen cow having a red coat are 82:246. If all of the cows in the herd have coats that are either black, red, or brown and if one cow is chosen at random from the herd, which coat color is the chosen cow most likely to have?

A. It depends on the actual number of cows.
B. There is a 1:3 probability of any of the colors.
C. Black
D. Brown
E. Red

Explanation of 25:

Notice what makes this question challenging. If the problem had said the odds of a black coat were 25:100 and the odds of a red coat were 35:100, you would know that out of 100 cows, 25 would have a black coat, 35 would have a red coat and the remaining 40 would have a brown coat.

This problem is more difficult because the odds are given in reference to a number more difficult to work with than 100 **and** because the odds for black use a different reference number than the odds for red.

To solve the problem we will need to express both odds in reference to the same number, like a common denominator. Before we attempt that, though, let's make this easier by looking at the answer choices to see what our options will be.

Choices A and B are interesting. Do you see why choice A cannot be correct? The odds have nothing to do with the number of cows, only with the ratio of characteristics. The odds are the same for 100 cows as for a million.

What about choice B? There are three colors. Why isn't there a 1:3 chance of each color? It is because the colors are not distributed equally. There is a greater predominance of some colors.

Now you have a 1:3 chance of picking the right answer by chance but we can do better than that. We need to express the odds using the same reference number so that we can compare them. This is exactly the same as having to add or subtract two fractions with different denominators.

$$\frac{81}{369} \text{ and } \frac{82}{246}$$

You already know how to do this with simpler numbers. If the numbers were

$$\frac{3}{5} \text{ and } \frac{2}{7}$$

you would multiply the first fraction by $\frac{7}{7}$ and the second fraction by $\frac{5}{5}$.

$$\frac{7}{7} * \frac{3}{5} = \frac{21}{35} \text{ and } \frac{5}{5} * \frac{2}{7} = \frac{10}{35}$$

However, in our case, multiplying by 369 and 246 would be cumbersome and prone to error.

You also probably know how to deal with fractions like

$$\frac{3}{5} \text{ and } \frac{2}{15}$$

In this case you can see that 15 is a multiple of 5, so you can create a common denominator just by multiplying $\frac{3}{5}$ times $\frac{3}{3}$.

In our problem 369 is not an integral multiple of 246. You might be able to figure out that 369 is exactly $1\frac{1}{2}$ times 246 but that is an intuitive insight that you might or might not catch.

If you do not see get that insight, there is a foolproof way to find the easiest common denominator. Break each denominator into its prime factors. We have worked on finding prime factors in recent problems. Here is a summary.

Start with 369. Consider the lowest prime number (not including 1). Is 369 divisible by 2? No. Go to the next prime. Is 369 divisible by 3? Yes. 3 goes into 369 exactly 123 times.

3				
369	123			

Now start the process of finding primes again from scratch with 123. The top row lists the prime factors that you have found so far. Is 123 divisible by 2? No. Is it divisible by 3? Yes, 41 times. Put 3 above it. To the right of it, in the bottom row, put 41.

3	3			
369	123	41		

You may remember a great tool for determining whether a number is divisible by 3. If you add the digits together and the result is divisible by 3, then the original number is also divisible by 3. Try it with 12,525. The digits add up to 15, which is divisible by 3. If you're not sure if 15 is divisible by 3, keep adding digits. 1 and 5 add up to 6. No matter how large a number is, if you add up the digits, then add up those digits and keep going until you have one digit, you will end up with 3, 6, or 9 if the original number is divisible by 3.

We still have to deal with 41. Is it divisible by 2? No. By 3? No. By 5? No. By 7? No. By 11? No. By 13? No. By 17? No. By 19? No. By 23? No.

When checking for prime factors, I can stop when I've gotten to a prime number that is just more than half of the number I'm checking. (There is a reason for this. See if you can figure it out.)

41 is not divisible by any of the prime numbers 23 or below. 41 must itself be prime. Enter it in the top row as the last prime factor.

3	3	41		
369	123	41		

We have now determined that the first ratio, 81:369 can be written as:

Chapter 4. ACT®-style Questions with Explanations

$$\frac{81}{3*3*41} \text{ or } \frac{3*3*3*3}{3*3*41} \text{ or } \frac{9}{41}$$

Now find the prime factors of 246.

2	3	41		
246	123	41		

Aha! 82:246 can be written as:

$$\frac{82}{2*3*41} \text{ or } \frac{2*41}{2*3*41} \text{ or } \frac{2}{6} \text{ or } \frac{1}{3}$$

Now, let's compare the two ratios for black and red coats:

$$\frac{9}{41} \text{ and } \frac{1}{3}$$

These two fractions are much easier to work with. The common denominator is 41*3. Multiply the first fraction by $\frac{3}{3}$ and the second by $\frac{41}{41}$.

$$\frac{3*9}{3*41} \text{ and } \frac{41}{3*41}$$

$$\frac{27}{123} \text{ and } \frac{41}{123}$$

Go back to the question stem and remind yourself what we are looking for. It is easy to lose track of that as you work through a question. We now know that out of 123 cows, 27 have black coats and 41 have red coats. How many remaining cows are there? 27 and 41 add up to 68. Subtract the 68 from 123 and there are 55 cows with brown coats. There are more cows with brown coats so it is more likely that a randomly chosen cow has a brown coat. Choice D is the correct answer.

If you are not comfortable with the additions and subtractions in the last paragraph and if you are not super accurate with a calculator, you can use a more intuitive way to find how many cows out of 123 have brown coats.

You need to subtract the 27 black cows and then the 41 red cows. Let's break those numbers into intuitive chunks. Using a table like the one below, from the 27 black cows first subtract 23 from 123 and put the remaining 4 in the middle column below the 23. You are choosing 23 because it is very easy to subtract from 123. This leaves 100 cows.

Starting number	Minus	Subtract	Remainder
123	27	-23	100
		-4	

Next bring the 100 remaining cows into the first column. Subtract the 4 that you had left over. The remainder is 96.

Starting number	Minus	Subtract	Remainder
123	27	-23	100
100		-4	96

Bring the 96 into the first column. You now need to start subtracting the 41 red cows. To keep things easy, start by subtracting 1 (from the 41) and put the rest of the 41 (40) into the middle column below it.

Starting number	Minus	Subtract	Remainder
123	27	-23	100
100		-4	96
96	41	-1	95
		-40	

You can now bring the remaining 95 into the first column. Subtract the 40.

Starting number	Minus	Subtract	Remainder
123	27	-23	100
100		-4	96
96	41	-1	95
95		-40	55

With the above method all of the subtractions are easy. You just need a good way to keep track of how much you have subtracted and what is left over.

The most intuitive process would be to draw 123 cows and cross them off as you remove 27 and then 41. But that would be a bit time consuming!

Such simple methods for more intuitive calculation might seem like overkill but many of the errors that people make on the test are due to incorrect calculation. It is a shame to get a question wrong that you have worked hard to understand only to make a calculation mistake.

You finished Day 5! How did it go on these five questions?

Number of questions you got right on your own: _____

Types of problems or patterns you need more work on: _____
_____

How much new did you learn from these questions? ☐ Important tools! ☐ Some tools ☐ Not too much

Day 6, Questions 26-30

26. At a charity event Glenda has a booth at which, without looking, she randomly draws three cards from a standard deck of 52 playing cards consisting of an equal number of black and red cards and places the cards face down on a counter. A participant then turns each card over. Each card is either red or black. The participant agrees to donate $5 for each red card. On the average how much money does a participant donate for one drawing of three cards?

F. $1.67
G. $2.50
H. $5
J. $7.50
K. $15

Explanation of 26:

The simplest way to look at this situation is just that, on the average, half of the cards have to be red. Cards can only be red or black. No matter how many cards are chosen, on the average half are red.

Because half of 3 is 1.5, on the average the participants will pay for 1.5 cards at $5 each, which comes to $7.50. Choice J is correct.

An intuitive way to calculate one and a half times $5 is to organize the information as below.

1 times $5	$5
Half of $5	$2.50
1 plus one half	$7.50

Chapter 4. ACT®-style Questions with Explanations

27. In the right triangle shown below, one of the angles measures $\tan^{-1}(\frac{1}{2})$. What is the $\sin[\tan^{-1}(\frac{1}{2})]$?

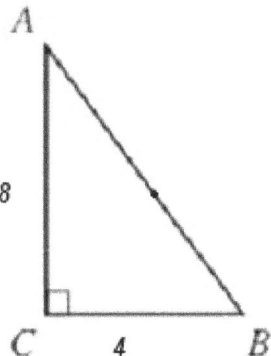

A. $\frac{4}{12}$

B. $4\sqrt{5}$

C. $\frac{8}{4}$

D. $8\sqrt{5}$

E. $\frac{1}{\sqrt{5}}$

Explanation of 27:

This is a great example of a question that is not as hard as it looks. You **do** need to understand what $\tan^{-1}(\frac{1}{2})$ means but it is not difficult. $\tan^{-1}(x)$ is called the inverse tangent but it is **not** the upside down of the tangent. The upside down of the tangent is the cotangent.

Tangent is opposite over adjacent. The cotangent is adjacent over opposite. What, then, is an inverse tangent? Whereas a tangent represents a ratio (opposite over adjacent) inverse tangent represents an angle.

If the tangent of $x°$ is $\frac{1}{2}$, then the inverse tangent of $\frac{1}{2}$ is $x°$. $\tan^{-1}(\frac{1}{2}) = x°$.

In other words the inverse tangent of $\frac{1}{2}$ is the angle that has a tangent of $\frac{1}{2}$. Inverse tangent refers to an angle. The diagram below illustrates this in an intuitive way.

Chapter 4. ACT®-style Questions with Explanations

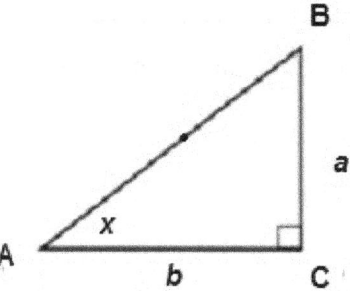

The tangent of x is $\frac{a}{b}$.

The $\tan^{-1}$ of $\frac{a}{b}$ is x.

If this is still a little fuzzy, it is similar to the difference between a grandchild and a grandparent. Ernest is the grandparent of Jordan. Jordan is the reverse grandparent of Ernest.

If all else fails, simply memorize that $\tan^{-1}$ refers to an angle.

Coming back to the problem, which angle has a tangent of $\frac{1}{2}$? It can be helpful to reorient the triangle as below.

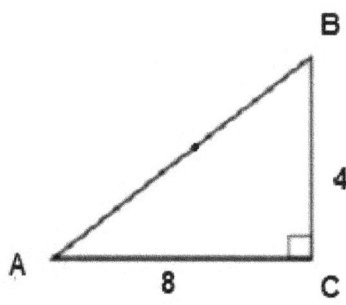

This is the standard orientation that we use to work with the angle at A. The tangent of angle A is $\frac{1}{2}$. Go back and remind yourself of what the question is asking. A lot of errors are due to getting this far and then forgetting what you are looking for. The question simply asks for the sine of angle A. You do **not** need to know the measure of angle A. However, you do need to know the length of the hypotenuse.

Use the Pythagorean Theorem. $4^2 + 8^2 = c^2$, where c is the hypotenuse.

$$c^2 = 16 + 64 = 80$$
$$c = \sqrt{80}$$

Because 80 is not a perfect square, the best tool for simplifying the square root is to break 80 into its prime factors.

$$2 * 40$$
$$2 * (2 * 20)$$
$$2 * 2 * (2 * 10)$$
$$2 * 2 * 2 * (2 * 5)$$
$$(2*2)(2*2)*5$$
$$4*4*5$$

Do you see that the square root of this is $4\sqrt{5}$? Prove it by squaring $4\sqrt{5}$:

$$4\sqrt{5} * 4\sqrt{5} = 4*4*\sqrt{5}*\sqrt{5} = 4*4*5$$

You have now determined that the hypotenuse is $4\sqrt{5}$. The sine is opposite over hypotenuse, or $\frac{4}{4\sqrt{5}}$ or simply $\frac{1}{\sqrt{5}}$. Choice E is the correct answer.

---

28. Given $f(x) = (x^2 + 5x)$, what is $f(2) * f(1)$?

F. 84
G. 20
H. 6
J. $x^4 + 10x^3 + 25x^2$
K. $\frac{x4 + 10x3 + 25x2}{2}$

Explanation of 28:

Look at the answer choices to orient yourself. The first three are specific numbers. The last two are expressions with variables. Which do you think the answer would be? Looking at the question stem, does it look like the answer will come out in specific numbers? You are going to plug 2 in for *x* and then plug 1 in for *x*. It seems that this should result in specific numbers. Choices J and K are traps that you might fall into if you made the problem more complex than it is.

Make a road map.

    Step 1. Calculate f(2).
    Step 2. Calculate f(1).
    Step 3. Multiply step 1 times step 2.

Step 1. $f(2) = (2^2 + 5*2) = (4 + 10) = 14$
Step 2. $f(1) = (1^2 + 5*1) = (1 + 5) = 6$

Chapter 4. ACT®-style Questions with Explanations

At this point you have to multiply 14 * 6. Before getting bogged down in calculations, compare choices F, G, and H. They are quite far apart. You only need to ask yourself whether 14*6 is larger than 20. It should be clear that it is. Choice F is correct.

If you wanted to use an intuitive approach for actually multiplying 14 times 6, you could regroup as below.

$$14 * 6$$
$$(2*7) * (2* 3)$$
$$(7 * 3) * 2 * 2$$
$$21 * 2 * 2$$
$$42 * 2$$
$$84$$

Remember to review the answer choices to see what you have to distinguish. Often you do not need to get all the way to a calculation to figure out the correct answer.

---

29. Sandy creates a magic trick in which a member of the audience is asked to come up with a positive two digit number N for which the tens digit is 3 greater than the ones digit. Then that member of the audience is asked to form a number P by reversing the digits in N and then must subtract P from N. Sandy is always able to tell the result. What is it?

A. -30
B. -3
C. 27
D. 30
E. 33

Explanation of 29:

There is an easy way and a hard way to solve this, though the hard way is more interesting.

Consider the situation carefully. If Sandy is always able to tell the result, that must mean that no matter which numbers the member of the audience chooses, the answer always comes out the same. All you have to do is to come up with one example and that must be the answer.

Follow Sandy's directions. Of the two digits, the ones digit can be considered the point of reference. In other words the tens digit is defined in terms of the ones digit. So let's start with a ones digit – 1.

The tens digit is defined as 3 greater than 1, or 4. Our number is 41.

The directions now tell us to reverse the digits – 14. Now subtract 14 from 41. If you are a bit shaky on this kind of subtraction, use a strategy that we discussed earlier. Subtract the 14 from 41 in small bits that are easy to subtract and use a table like the one below to keep track of each step.

Starting number	Minus	Subtract	Remainder
41	14	-10	31
31		-1	30
30		-3	27

The answer is 27, so choice C is correct.

Now, if you are interested, we can look at why this works. If *d* represents the ones digit, then the tens digit, *t*, is $d + 3$. Consider that a two digit number *td* equals $10*t$ plus $1*d$. So our number is $10(d + 3) + d$ or $10d + 30 + d$ or $(11d + 30)$.

When we reverse the digits, (*dt*) we get $10d + (d + 3)$, because the units digit is now $d + 3$ (=*t*). Simplify this to $11d + 3$.

Now subtract the reversed number from the original number:

$$11d + 30 - (11d + 3)$$
$$11d + 30 + -1(11d + 3)$$
$$11d + 30 + -11d + -3$$
$$11d + -11d + 30 + -3$$

Notice that, at this point, the *d*'s drop out. This is why it does not matter what numbers the audience member chooses. What is left is $30 - 3$ or 27.

Also notice that instead of dealing with subtraction in the expressions above, in the second row we changed the subtraction to addition and multiplied the expression in parentheses by -1. Additions are less prone to errors than subtractions.

30. On a map of his state Edgar measures the distance between his hometown and the state capital as 2.5 inches. If the distance between these two towns is 210 kilometers, how many inches on the map will represent a distance of 300 kilometers, to the nearest tenth of an inch?

F. 2.8
G. 3.0
H. 3.6
J. 3.8
K. 4.2

Explanation of 30:

This problem deals with a single ratio, 2.5:210. This same ratio can be expressed in various ways. For example the ratio 5:420 is the same ratio. Both sides of the ratio have simply doubled. The question is asking us to find *x* where *x*:300 is the same ratio as 2.5:210.

Let's use a table to explore how to find equivalent values of the ratio using the intuitive strategy of creating building blocks.

2.5	210
?	300

Starting with the ratio that we know, we can double it, halve it or any other manipulation that is intuitively easy, in order to create new building blocks. Once we have several building blocks, we can add or subtract them to get other blocks. We want to get close to 300 so let's create a building block for half of the original ratio, namely 1.25:105.

1.25	105
2.5	210
?	300

We can now add the values for 1.25 and 2.5. That gives 3.75 inches and 315 kilometers.

1.25	105
2.5	210
?	300
3.75	315

This is useful information. Check the answer choices. We know that the number of inches that corresponds to 300 **must** be lower than 3.75. Choices J and K are out. The value for 300 is only a little bit lower than 3.75, so choice H looks good but we need to confirm that. We can test the answer choices. Testing 3.6 would be a little complicated. Testing 3.0 (choice G) might be easier.

To get to 3.0 inches we could add 0.5 to 2.5. We can get the value for 0.5 by dividing the ratio for 2.5 by 5. 210 divided by 5 is 42, as shown in the table below. Adding the values for 0.5 and 2.5 gives a value for 3 of 252 kilometers.

0.5	42
1.25	105
2.5	210
3.0	252
?	300
3.75	315

This shows that 3.0 is too small to be the correct answer and choice H must be correct.

This method of creating various versions of the same ratio is very powerful.

You finished Day 6! How did it go on these five questions?

Number of questions you got right on your own: _____

Types of problems or patterns you need more work on: _____

_____

How much new did you learn from these questions? ☐ Important tools! ☐ Some tools ☐ Not too much

Day 7, Questions 31-35

31. Given the equation $\log_x 81 = 4$, which of the following is closest to the value of x?

A. 3
B. 3.25
C. 4
D. 9
E. 20.25

Chapter 4. ACT®-style Questions with Explanations

Explanation of 31:

Solving this problem requires being very clear on the logarithm terminology. Here is a simple review.

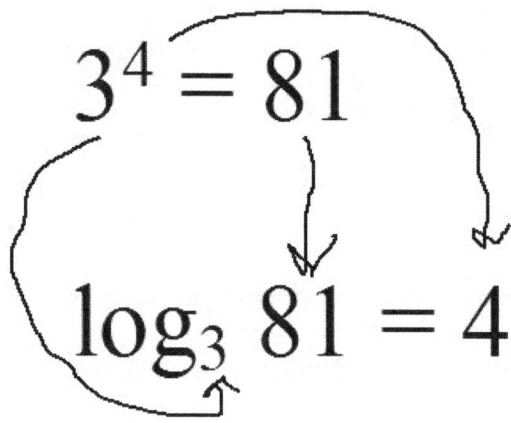

In the diagram above you can compare the expression $3^4 = 81$ with its equivalent in terms of logarithms. Follow the arrow for each of the three elements in the top expression to see where it goes in the bottom expression.

3 is the base of the log. That means it is the base for the exponent to be attached to. In other words, we are talking about powers of 3.

Log_3 is another way to say we are dealing with 3 to some power. The expression $log_3 81$ means we are talking about 3 to some power that will equal 81. The equation $log_3 81 = 4$ tells you that 3 to the fourth power equals 81.

Another way to picture this relationship is:

Log_{Base} BigNumber = Exponent

This means that the Base taken to the Exponent Power gives the BigNumber. You can also remember that a log always points to an exponent. The log of some number is an exponent. The exponent goes on the base.

Yet another way to think of logs is that a log means an exponent. "What it the log of 25" means what is the exponent that gives you 25. Of course it depends on what the base number is. If the base is 5, then the exponent would be 2. $5^2 = 25$. If the base was a different number, the exponent would be different, so when a question asks you what the log (exponent) of a number is, it also has to tell you what base it is talking about. "What is the exponent (log) that gives 25 if the base is 5?" This is the same as saying $log_5 25 = 2$.

For some people this is a difficult relationship to keep straight. If you work with it a bit you might find other ways to remember this clearly.

Apply this to the original question. $\log_x 81 = 4$. What is x? First, you know that x is the base. It has to be raised to a power and that power is 4. x raised to the fourth power gives the "BigNumber", 81.

There are a number of ways to solve for x now. You can test answer choices. To do this, stick with choices that are easy to test. What about choice A, 3? Is $3^4 = 81$? Actually, it is. $3*3 = 9$ and $9*9 = 81$. We got lucky. Choice A is correct.

Another way to look at a problem like this if testing answers does not work is to consider that we are looking for:

$x * x * x * x = 81$

In essence, x is the fourth root of 81. Finding the prime factors of 81 may reveal the fourth root. 3 goes into (81) 27 times. 3 goes into (27) 9 times. 3 goes into (9) 3 times. 3*3*3*3.

Another approach to find the fourth root is to take the square root of the square root. The square root of 81 is 9. The square root of 9 is 3.

32. The figure below is a trapezoid. The length of $\overline{AB}$ is 8.

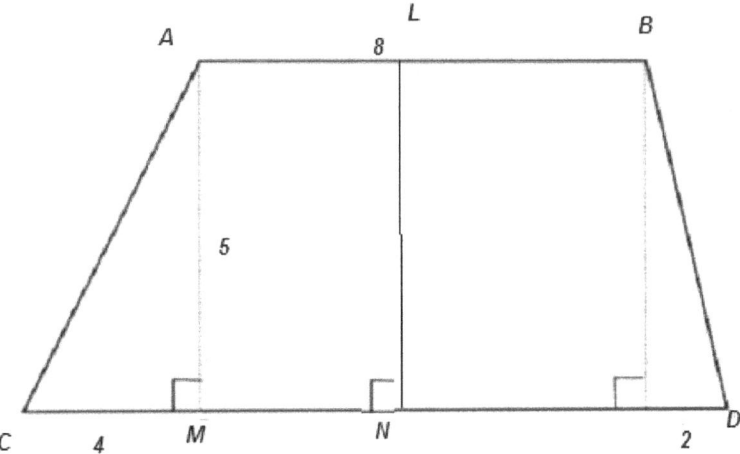

(*Figure not drawn to scale.*)

If a perpendicular vertical line $\overline{LN}$ is inserted such that it bisects the trapezoid into two parts with equal area, what is the length of $\overline{AL}$?

F. 3.5
G. 4
H. 4.5
J. 17.5
K. 27.5

Chapter 4. ACT®-style Questions with Explanations 99

Explanation of 32:

This is a complex figure. If the top of the trapezoid were centered over the base of the trapezoid, it would be easy to find the middle line of the entire trapezoid. In this question the top is offset.

The best way to approach this is to divide the figure up into smaller, more regular figures. The diagram shows the figure broken up into triangle CMA on the right and a smaller triangle on the right with points B and D and an unnamed point. In addition to the two triangles, there is a rectangle with corners A, M, B and the unnamed point.

Check to see if we have enough information to calculate the areas of these three regions. The left triangle shows the base and the height, so we can calculate the area. The right triangle shows the base. The height must be the same as $\overline{AM}$, 5, because the trapezoid is "right", having right angles. The rectangle has one side of 8 and one side of 5. There is enough information to calculate the areas of all three regions.

Review the question stem to remind yourself of what the question is. The question is not asking for the total area or half the area. It is asking for the position of the line that divides the figure in half, meaning that both parts have the same area. How can you get to that?

From an intuitive standpoint, there are at least two ways to look at this. Let's consider them. You can calculate the area of the two triangles and determine how much larger one is than the other. Then you need to divide the rectangle into two unequal pieces to make up for the difference in area between the triangles, as shown in the figure below.

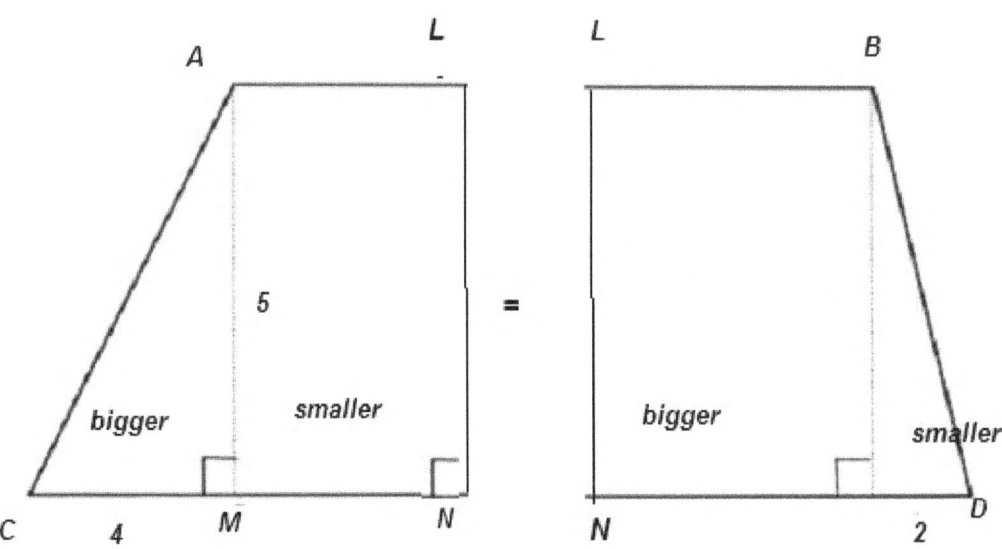

Another way to solve the problem could be to calculate the area of the entire trapezoid, determine what half of that area is, and use that information to figure out where $\overline{LN}$ needs to be drawn.

When you are faced with two different options for pursuing a solution, it is hard to know which to try first. If one intuitively feels simpler, go with that one. If your first attempt bogs down, switch to the other approach.

Here we have the luxury of trying both in order to build intuitive problem solving skills. We can start with the first approach. The areas of the two triangles are easy to find, using area = $\frac{1}{2}$bh. The left-side triangle has an area of 10. The right-side triangle has an area of 5. The area of the rectangle is 8*5, or 40.

Review Area of Triangles

If you're a little rusty on the formula for the area of a triangle, here is a good way to remember it.

Consider a rectangle divided into two equal parts by a diagonal line.

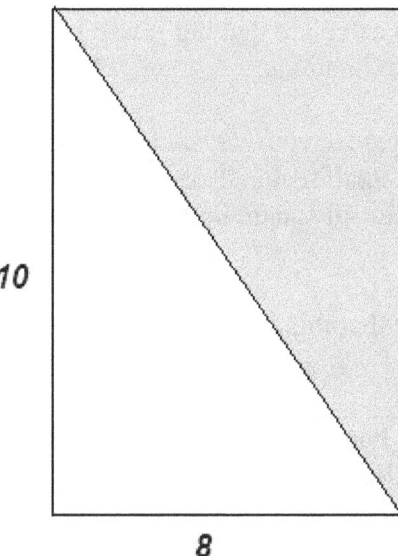

The area of the rectangle is the width times the length, which we can also call the base (8) times the height (10). The area is 80.

What is the area of the unshaded triangle? Isn't it simply half of the area of the rectangle, or $\frac{1}{2}$bh? This formula for the area of a triangle is true for any triangle, not just right triangles.

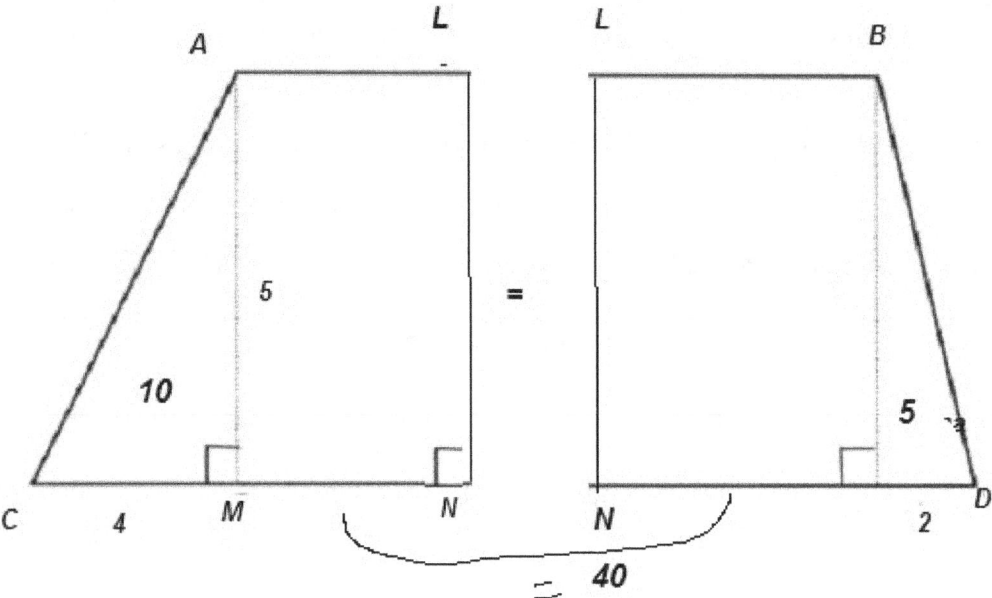

Your job now is to divide up the 40 square units of the rectangle in such a way that the two figures have equal area. There is an algebraic way to calculate that but it will be more fool-proof to use an intuitive approach, based on trying various approximations.

Use a table such as the one below to organize your work in a way that makes sense to you. Let's try an approximation. First, is it clear that the small rectangle on the left will need to be smaller than the small rectangle on the right? Let's divide up the 40 square units into 15 on the left and 25 on the right as a first approximation.

You can see in the first row of the table that this doesn't result in equal areas. Let's try 18 and 22 as an approximation.

Triangle on left	Small rectangle	Total for left	Triangle on right	Small rectangle	Total for right
10	15	25	5	25	30
10	18	28	5	22	27

This is closer. Here is an important point to notice. In the first attempt, the right hand figure was too large. In the second approximation, the left hand figure is too large. This means we have already passed the number that we need! The number that works is between 15 and 18 for the left hand side and closer to 18.

It might take several more approximations to get the correct numbers. So far we have tried whole numbers, which are easier to work with. As we get closer, we might need to test mixed numbers. Let's try 17.5 and 22.5 (=40).

102 Chapter 4. ACT®-style Questions with Explanations

Triangle on left	Small rectangle	Total for left	Triangle on right	Small rectangle	Total for right
10	15	25	5	25	30
10	18	28	5	22	27
10	17.5	27.5	5	22.5	27.5

Bingo. If we had tried 17 and 23, we would have found that the 17 was still too small.

Note that we are still not at an answer. We've determined what the two equal parts of the trapezoid look like but we need to determine where L is.

Let's put that aside for now and consider the other approach for dividing the trapezoid into two equal pieces. The area of the entire trapezoid is 10 + 5 + 40 = 55(the left hand triangle plus the right hand triangle plus the remaining rectangle. This means that the area of each half must be 27.5. (Does that number look familiar??)

In the figure below we know the areas of each triangle. It is then simple to calculate what the area of the rectangle must be for the entire half to have an area of 27.5.

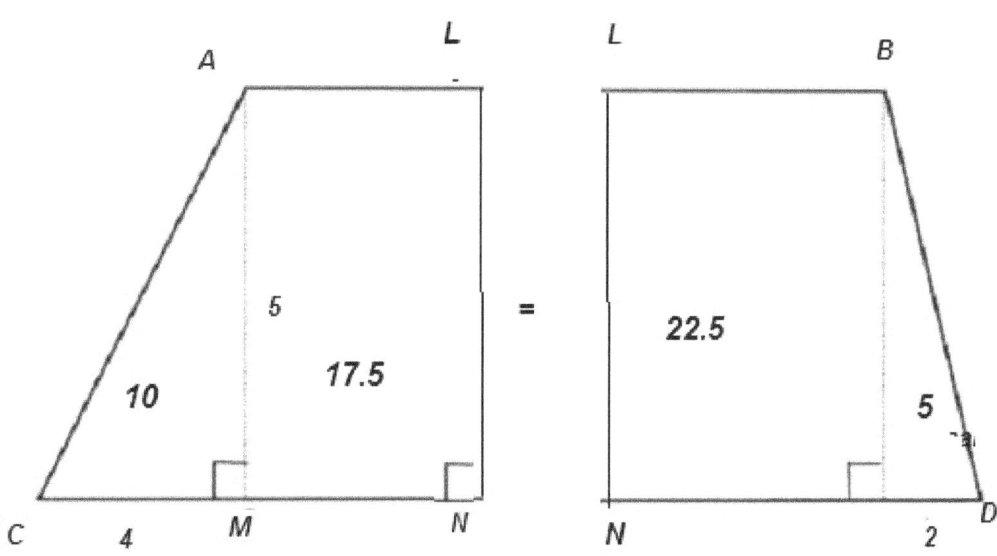

Which method seemed the easiest to you?

Now that we know what the two equal halves look like, go back to the question stem again to remind yourself of what we are looking for. We need to know the length of $\overline{AL}$. How can you calculate that?

We know the area of rectangle ALNM. We know one side. $5 * \overline{AL} = 17.5$.

Solve for $\overline{AL}$. But don't do more work than you need to! Check the answer choices to see how close they are. Approximate $\overline{AL}$. It has to be a little more than 3, right? The answer choices are 3.5, 4, 4.5 and up. Could $\overline{AL}$ be as much as 4? 5*4 = 20 and that is greater than the area of ALNM.

The correct answer must be less than 4 and greater than 3. Choice F, 3.5, is the only answer that fits. Using this approach avoids the possibility of making a calculation error and is probably faster.

---

33. Last month Sheri earned $145 doing freelance programming. In reviewing how much she earned this month, Sheri calculated that she earned 7% more than she had last month. How much did Sheri earn this month?

A. $134.85
B. $147.70
C. $152.70
D. $155.15
E. $159.00

Explanation of 33:

This will not be too tricky if you orient yourself to the problem carefully. Let's use some intuitive strategies to estimate the answer. If she had earned $100 last month, 7% more would be exactly $7. Because she earned more than $100, the increase must be more than $7. Can you eliminate any answer choices?

The original $145 plus $7 would be $152. The correct answer must be larger than that. Choices A and B are out. Choice C looks too close to $152 but you will need to test it.

Let's do more intuitive estimation. If Sheri had earned $150, how much would 7% more be? 150 = 100 + 50. 7% of 100 is $7. $50 is half of $100 so 7% of 50 is half of $7, or $3.50. Added together, 7% of $150 is $10.50. If we add that to the original $145, we get $155.50.

If it was not easy to follow the above paragraph, you can do the same work with a table.

Amount	7%
100	7.00
50	3.50
150	10.5

$155.50 is based on Sheri earning $150 last month but she only earned 145. The correct answer must be less than $155.50. Choice E is out. Choice D looks very close and choice C looked to small. It would be tempting to go with choice D (and you would be correct) but let's use intuitive tools to make sure.

Using the table below, we can break our ratio up into small units in order to find exactly what 7% added onto $145 would be. First we will calculate the rate for $10 by dividing the amount for $100 by 10. The table shows that it is $0.70. Next we can divide that number in half to get the rate for $5. Using groups of $10 and $5, let's build up to 145. Instead of multiplying the amount for $10 by 4, let's first double it and then double it again. Doubling is more foolproof and more intuitive than multiplying by four. The final row in the table shows the amounts for $145.

Amount	7%
100	7.00
10	0.70
5	0.35
20	1.40
40	2.80
100 + 40 + 5	7.00 + 2.80 + 0.35

If you are a little iffy about accurately adding the three numbers in the bottom right column, use an intuitive method. You can add the 2.80 to 7.00 intuitively. Then break down the 0.35 into parts that you can easily add.

7.00	2.80	9.80
9.80	0.10	9.90
9.90	0.10	10.00
10.00	0.15	10.15

Adding the $10.15 onto the original 145 gives $155.15. This confirms that choice D is the correct answer.

If you got a little lost in the explanation above, go back to the beginning of it and create a road map for yourself. Then check where we are on the road map as you read the explanation.

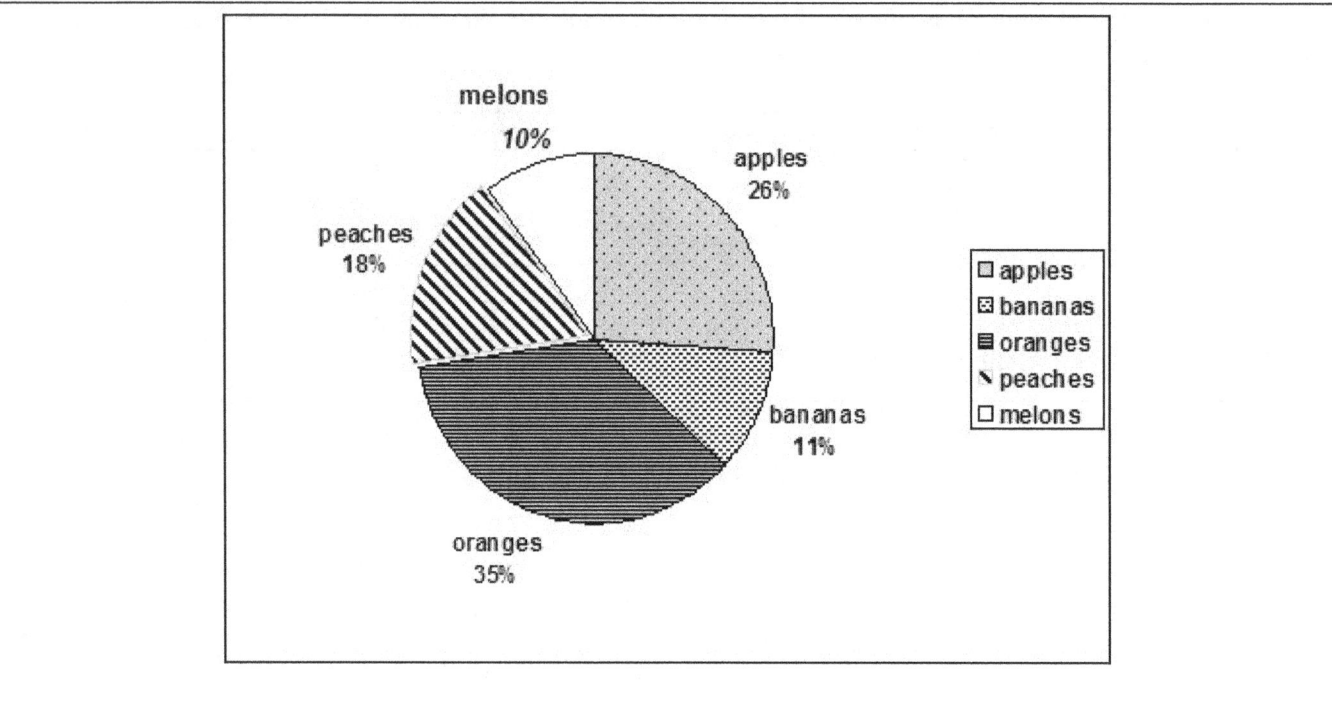

34. The circle graph above shows the percent of all fruits carried by a certain small grocery store. If these are the only fruits sold in the store and if at a particular time there are exactly 275 pieces of fruit for sale, what are the odds (chosen type of fruit:not chosen type of fruit) that a randomly chosen piece of fruit will be an apple?

F. 68:275
G. 26:275
H. 1:68
J. 13:37
K. 13:50

Explanation of 34:

There are a couple tricks in this question. The first is that this question is **not** the same as asking what the odds are that a randomly picked piece of fruit is an apple. The usual odds questions would be expressed as the number of apples compared to the total number of fruits. This question is asking for the odds expressed as number of possibilities for an apple compared to number of possibilities for non-apple fruits.

The second trick is that you may be tempted to figure out how many apples there are (26% of 275) and how many non-apple fruits there are. This is not necessary. Here's why.

The answer choice is a ratio, not a specific number. As long as the ratio of fruits to each other stays the same, the ratio of apples to non-apples is the same whether there are 100 total fruits or 1,000,000 total fruits.

To get the answer, you only have to determine that for every 100 fruits, 26 are apples and the remaining 74 are not apples. The ratio of 26 to 74 can be reduced to 13:37. Choice J is the correct answer.

---

35. Given the system of equations below, for what value of m would the system have an infinite number of solutions?

$$2x^2 - 3y = 20$$
$$10x^2 - 15y = 10^m$$

A. 8
B. 2
C. 1
D. 0
E. -2

Explanation of 35:

This is another trick question. It is always a good idea to start a problem by orienting yourself carefully to the question stem. Ask yourself if it makes sense. Identify things that might be confusing. Clarify terms.

In this case you can ask yourself how two equations can have an infinite number of solutions. At first this does not seem to make sense. Let's break it down.

For one equation with two variables, such as x and y, how many solutions are there? In most cases there are an infinite number. For example, $x + y = 7$. The x and y can be 1 and 6, 2 and 5, 3 and 4, 4 and 3, 1.5 and 5.5, and so on. There are an infinite number of pairs.

However, when a second equation with x and y are added, the number of solutions is limited. This problem gives two equations with x and y, so how can there be an infinite number of solutions? Consider the system of equations below.

$$x + y = 7$$
$$y + x = 7$$

Even though there are two equations, there are still an infinite number of solutions. Why? Because there are not two **different** equations. The two equations are the same. What about:

$$x + y = 7$$
$$2x + 2y = 14$$

These are still the same two equations. The second one can be reduced to the first.

The only time that you can have two equations in two variables and still have an infinite number of solutions is if the equations are the same (equivalent.)

Examine the two equations in the problem carefully. The left side of the second equation is must 5 times the left side of the first equation. If the right side of the second equation is also 5 times the right side of the first, there will be infinite solutions.

$$5 * 20 = 10^m$$
$$100 = 10^m$$
$$100 = 10^2$$

Therefore, $m = 2$. Choice B is correct.

You finished Day 7! How did it go on these five questions?

Number of questions you got right on your own: _____

Types of problems or patterns you need more work on: _____

_____

How much new did you learn from these questions? ☐ Important tools! ☐ Some tools ☐ Not too much

Day 8, Questions 36-40

36. Andy makes two kinds of candy – chocolate truffles and peanut clusters. The amount of money in dollars that he makes from selling the candies is expressed as $15t + 7p$, where t is the number of truffles and p is the number of peanut clusters. Andy's cost for making one truffle is $1 and his cost for making a peanut cluster is $0.40. In a given week Andy sold 5 truffles and 10 peanut clusters. His cost for producing those candies was approximately what percent of his income from selling them?

F. 3%
G. 6%
H. 9%
J. 10%
K. 12%

Explanation of 36:

This problem has a lot of steps. You need to find Andy's cost for producing the candies he sold and you need to find the income that he made from them. You then need to put the cost over the income and convert that ratio to a percent.

$$\frac{\cos tOf \Pr oduction}{income} = \frac{?}{100}$$

This is a good time to create a road map!

Step 1. Calculate the cost of producing the candies sold.
Step 2. Calculate the income from the candies sold.
Step 3. Calculate step 1 over step 2.
Step 4. Convert step 3 into a percent.

Step 1. Organize the information intuitively and visually. This makes it easy to understand and prevents confusion and mistakes.

Item	#	Cost per item	Total
truffle	5	1	5
peanut cluster	10	0.40	4
Total			$9

The candies cost Andy $9.

Step 2. The formula for his income is $15t + 7p$. Plug in the numbers.

$$15 * 5 + 7 * 10$$
$$75 + 70$$
$$\$145$$

> ### Intuitive Multiplication
>
> A more intuitive way to multiply 15 times 5 is to break 15 into prime factors and see if you can regroup them in a way that makes it easy for you to do the multiplication.
>
> $$3 * 5 * 5$$
>
> Regrouping the two 5's together gives $3 * 25$, which you may find more intuitively easy to multiply. $3 * 25 = 75$.

Step 3. $\dfrac{9}{145}$

Step 4. The ratio in step 3 is not particularly easy to reduce. If you are comfortable with your accuracy on the calculator, you can use that where allowed. Otherwise, you can use intuitive tools.

Let's first use estimation and approximation. $\dfrac{9}{145}$ is smaller than $\dfrac{9}{100}$, which is 9%. Any answer that is 9% or more can be eliminated. Choices H, J, and K are out.

Compare the remaining choices. It is hard to tell at a glance which seems most likely. We can test out one of them. Let's test 3%. Is 9 3% of 145? Consider the table below. You can calculate 1% of 145 easily. Then you can double it. Adding the two gives you 3%. It's more foolproof to double and to add than to multiply by 3.

You don't need to get the decimals exactly correct to see that this number is not close to 9. The other remaining answer choice, 6%, is double 3% and puts you in the right range. Notice that the question asks for the approximate percent. The exact percent is 6.2069, in case you were wondering. Choice G is the correct answer.

1%	1.45
2%	2.9
3%	1.45 + 2.9 = 4.35

37. For every 15 students in a beginning programming class, a certain school district receives a $270 donation from a local software company. If this semester the class has 25 students and if the software company bases its donation on the above ratio, how much does the software company donate this semester?

A. $350
B. $375
C. $400
D. $425
E. $450

Explanation of 37:

This is a straightforward ratio question. The ratio is 15:270. We need to know what the total would be for 25 students instead of 15. Using a table like the one below, first enter the basic ratio and the target.

15	270
25	

Our goal is to break 15 down in intuitively easy ways to create building blocks so that we can get to 25. The easiest way to divide a number is by 2 or by 10 but in this case that does not help. Instead, both 15 and 270 can be divided by 3.

Is a Number Divisible By 3?

Do you remember the strategy of adding the digits of a number to see if it divisible by 3? For 270, 2 + 7 = 9. 9 is divisible by 3, so 270 is divisible by 3.

On the top line, enter 5 and 90. Now we can use 5 as a building block to get to 25. 5 times 5 is 25 and 5 * 90 is 450. The answer must be 450.

110   Chapter 4. ACT®-style Questions with Explanations

5	90
15	270
5 * 5 = 25	5 * 90 = 450

If you are not very comfortable with multiplying 5 times 90, you can break information in the table down a little more. Create an entry for 10 on the second line, by doubling the line for 5. Now create a line for 20 by doubling the line for 10. Now you can add the lines for 20 and for 5 to get 25 and 450. This method keeps the addition a little simpler.

5	90
10	180
15	270
20	360
5 + 20 = 25	360 + 90 = 450

Choice E is the correct answer. Notice that when we created the tables above, we included extra lines to be able to fill in new building blocks as needed in their correct places.

---

38. If the number of bacteria in a colony at a given time t can be expressed as $t^{\frac{3}{5}}$, which of the following expressions would always give the same result for the number of bacteria?

F. $\dfrac{3}{5}t^2$

G. $\dfrac{t^3}{5}$

H. $\dfrac{1}{t^{\frac{5}{3}}}$

J. $\sqrt[5]{t^3}$

K. $\sqrt[3]{t^5}$

Explanation of 38:

Consider what the question is asking. The good news is that you are not being asked to calculate the number of bacteria in some situation. Is it clear that the question is asking you to find an expression that is equivalent to the original?

The task here is to understand what a fractional exponent means. Let's break this into two parts - t^3 and $t^{\frac{1}{5}}$. t^3 is easy to understand. It is simply t raised to the third power. You can eliminate any answer choice that does not include t to the third. Choices F and K are clearly out. Choice H looks dubious but probably should not be eliminated.

Now consider $t^{\frac{1}{5}}$. Many people confuse this with t^{-5}. Let's look at a way to understand these two expressions and be able to distinguish them.

First, there are only two things that the above expressions could mean. One is the fifth root of t and the other is $\frac{1}{t^5}$. Which is which?

You might need to simply memorize that the one with the fraction in the exponent does **not** lead to the fraction in the result. In other words

$$t^{\frac{1}{5}} = \sqrt[5]{t}\text{, and}$$

$$t^{-5} = \frac{1}{t^5}$$

Play around with these expressions and see if you can make up a way to keep them straight. What's important is that you find a way that works for you.

Here's a story. Maybe it will help.

Cyd was walking through the woods and had $10. First, Cyd ran into a good magician, who said, "I'm going to raise your money to the power of 2." Suddenly, Cyd had $100. Next Cyd ran into a not so good magician who said, "I'm going to raise your money to the power of $\frac{1}{2}$. Suddenly, Cyd was back to $10.

Finally, Cyd ran into a really mean magician, who said, "I'm going to raise your money to the power of – 2. Poor Cyd was left with one cent.

Or maybe this will help.

10^2	$10*10$	100
$10^{\frac{1}{2}}$	$\sqrt{100}$	10
10^{-2}	$\frac{1}{10^2}$	0.01

In the table above, the rows are in diminishing order. The fractional exponent generally gives a larger result than the negative exponent. In other words, the exponent 2 gives a larger result than the exponent $\frac{1}{2}$, and the negative exponent, -2, gives the smallest result of all, $\frac{1}{10^2}$.

Remember: the fraction does **not** lead to the fraction.

Coming back to the problem, what does $t^{\frac{1}{5}}$ represent? The fraction does **not** lead to the fraction. This must mean the fifth root of t, $\sqrt[5]{t}$. Of the remaining answer choices, only choice J contains a fifth root. That is enough information to choose choice J as the correct answer.

---

39. At Cidney's Diner 35% of the patrons order the turkey dinner and 25% order the enchilada plate. If 49 people order the turkey dinner, approximately how many order the enchilada plate?

A. 12
B. 25
C. 35
D. 49
E. 50

Explanation of 39:

In this problem you have to consider ratios and actual numbers. We can use a familiar intuitive method. Start by entering the information that we have to match a percent with an actual number, namely that 35% equals 49 people. Enter it on the last line of the table below. Then enter our target percent, 25%, on the row above it. Then create a smaller building block by dividing 35 by an intuitive number. Both 35 and 49 can be divided by 7, so create a row for 5% and 7 people.

5%	7
25%	
35%	49

Use 5% as a building block to build rows for 10% and then 20%. This only involves doubling, which is intuitively easy. You can now create a row for 25% by adding the rows for 20% and 5%.

5%	7
10%	14
20%	28
5% + 20% = 25%	7 + 28 = 35
35%	49

The correct answer is choice C, 35. You got there without any algebra or complex multiplications, cross multiplications or calculator work.

40. Of Tranh's friends, 15 report being vegetarians and 13 report being joggers. If Tranh has 29 friends, what is the minimum number who must be both vegetarians and joggers?

F. 0
G. 1
H. 13
J. 15
K. 28

Explanation of 40:

This is a trick question but you can avoid the trap if you look carefully at the answer choices. The first choice is zero. Could that be the case? The vegetarians and the joggers add up to 28. Tranh has 29 friends. It is possible that there is no overlap at all.

The question asks for the smallest possible overlap. For such a question you should consider the smallest answer first. If the smallest answer does not work, check the next smallest one.

You finished Day 8! How did it go on these five questions?

Number of questions you got right on your own: _____

Types of problems or patterns you need more work on: _____

_____

How much new did you learn from these questions? ☐ Important tools! ☐ Some tools ☐ Not too much

Day 9, Questions 41-45

41. For $-3 < x < -2$, which of the following represents possible positions of x^2?

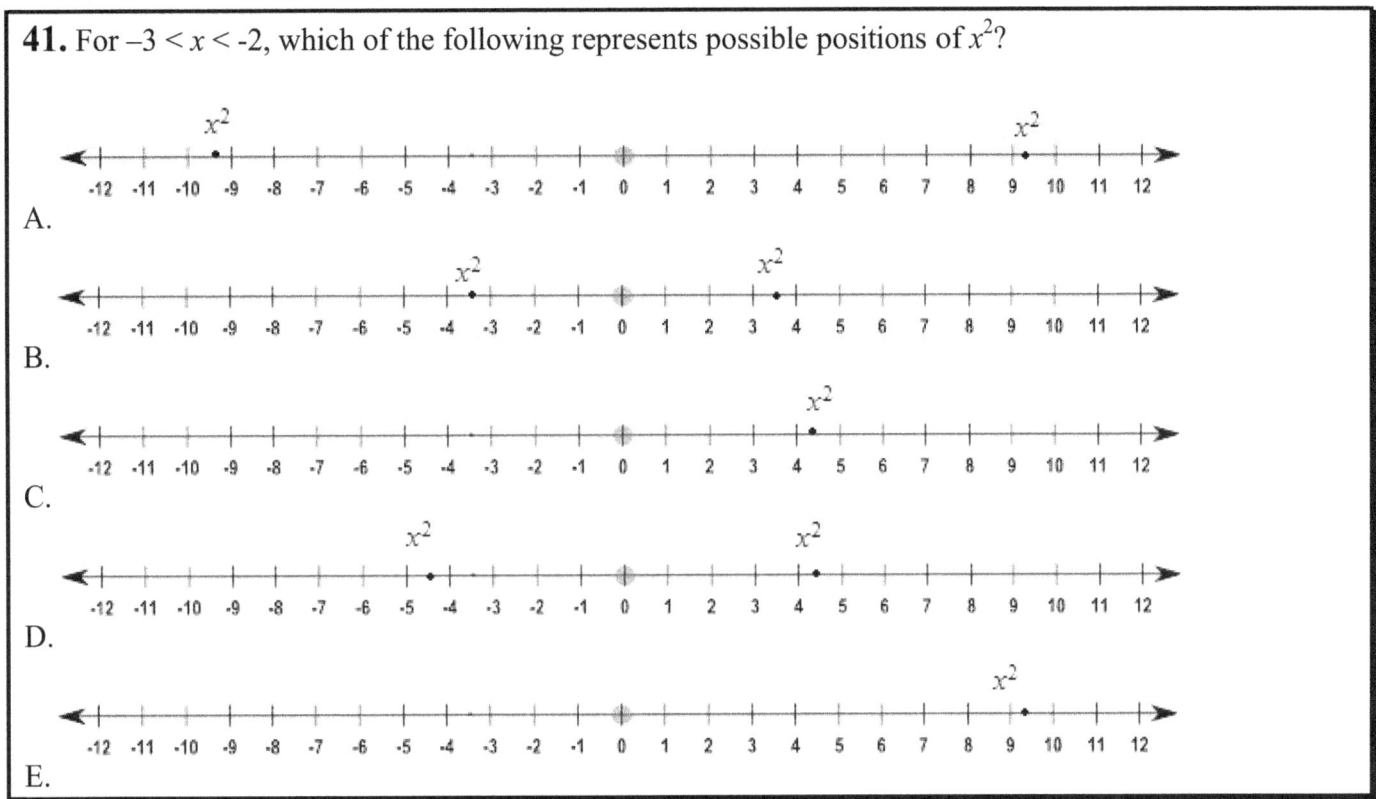

Explanation of 41:

Orient yourself carefully to this question. What do you know about the answer choices that are **incorrect**? They include a position for x^2 that is not possible. This is important information to be aware of. It may be easiest to eliminate answer choices because they contain an incorrect position.

Consider what you know about x and x^2. There are two elements. x is negative and x is between -3 and -2. Let's deal with these separately.

First, the fact that x is negative does not affect the fact that x^2 must be positive. Any answer choice with a negative value for x^2 can be eliminated. Choices A, B, and D are out.

Only choices C and E are left. Comparing two answer choices is a powerful intuitive strategy. It allows you to focus on the differences so you can distinguish the correct answer.

Choice C puts x^2 between 4 and 5. Choice E puts x^2 between 9 and 10. Which is correct? We do not know exactly where x is but we do know its limits. It is between -2 and -3. We can calculate the limits for x^2. They are 4 and 9. Choice C falls within those limits. Choice E does not. The correct answer is choice C.

42. On the standard (x,y) coordinate plane below, the *x-axis* represents the number of incidents of an event occurring and the *y*-axis represents the number of incidents of the same event not occurring. The overall probability of the event occurring is represented by $\sqrt{x^2+y^2}$. Which of the five points shown on the plane has the greatest probability of occurring?

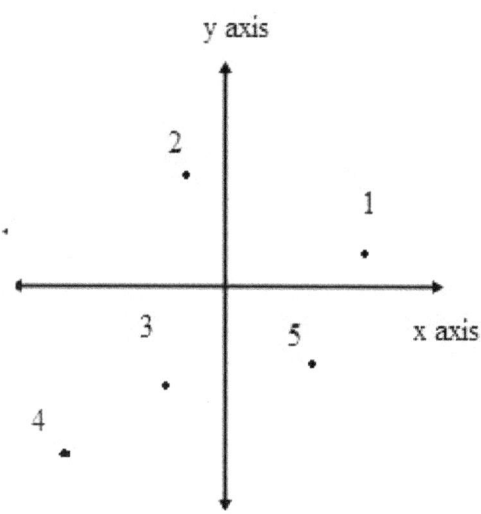

F. 1
G. 2
H. 3
J. 4
K. 5

Explanation of 42:

The problem asks you to find the point for which the value of $\sqrt{x^2+y^2}$ is the largest. You can actually ignore the square root, because the largest square root will be the square root of the largest x^2+y^2.

There are no units marked on the diagram but you can arbitrarily assign units to get a rough idea of the values of *x* and *y* for each point. Suppose we say point 1 is at (5,2). Rough approximations of the other points, in numerical order, would be (-2,5), (-3,-3), (-5,-5), and (3,-2).

Combining the squares of each gives, in numerical order, (25 + 4), (4 + 25), (9 + 9), (25 + 25) and (9 + 4). The fourth point clearly gives the largest number.

As an interesting side note, the expression $\sqrt{x^2+y^2}$ probably reminds you of the value of the hypotenuse in a right triangle. This is not a coincidence. Consider point 4 as the apex of a triangle, as shown below.

116   Chapter 4. ACT®-style Questions with Explanations

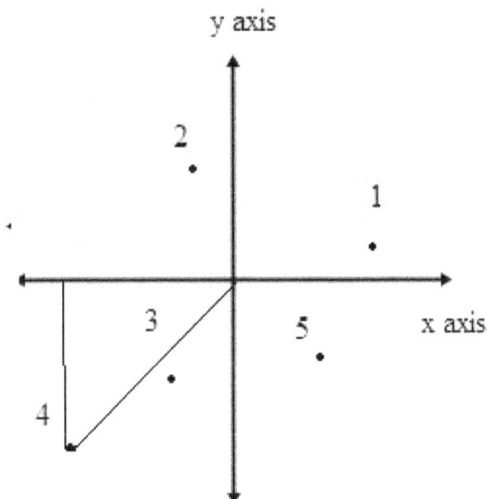

The length of the hypotenuse, which is the segment from point 4 to the origin, is exactly $\sqrt{x^2 + y^2}$. To find the point with the greatest value of $\sqrt{x^2 + y^2}$, you could simply measure the distance from the point to the origin.

43. In each of the past five years Ted spent the following amounts respectively on vacations:

| $1220 |
| $1290 |
| $1240 |
| $1220 |
| $1275 |

What is the mean dollar amount that Ted spent on vacation over these five years?

A. 1240
B. 1245
C. 1249
D. 1250
E. 1255

Explanation of 43:

Remind yourself that "mean" means average. This is not the same as the median. The standard way to get the average is to add all of the amounts and then divide by the number of items. However, this leaves a lot of room for making a calculation error.

There are some intuitive short cuts. For one thing, all the numbers are in the 1200's. You can ignore the leftmost two digits and simply average 20, 90, 40, 20, and 75.

Chapter 4. ACT®-style Questions with Explanations

Even for doing that there are some intuitive tools that improve your accuracy. Let's test an answer choice. It may not turn out to be the correct answer but it may help us focus in on the correct answer. Start with 1250 (or just 50) because it is so easy to work with.

The diagram below is a powerful tool for working with averages. The horizontal line through the middle represents the average. Each point falls on, above, or below the average. By keeping track of how far above or below, you can make conclusions about the average of the five points.

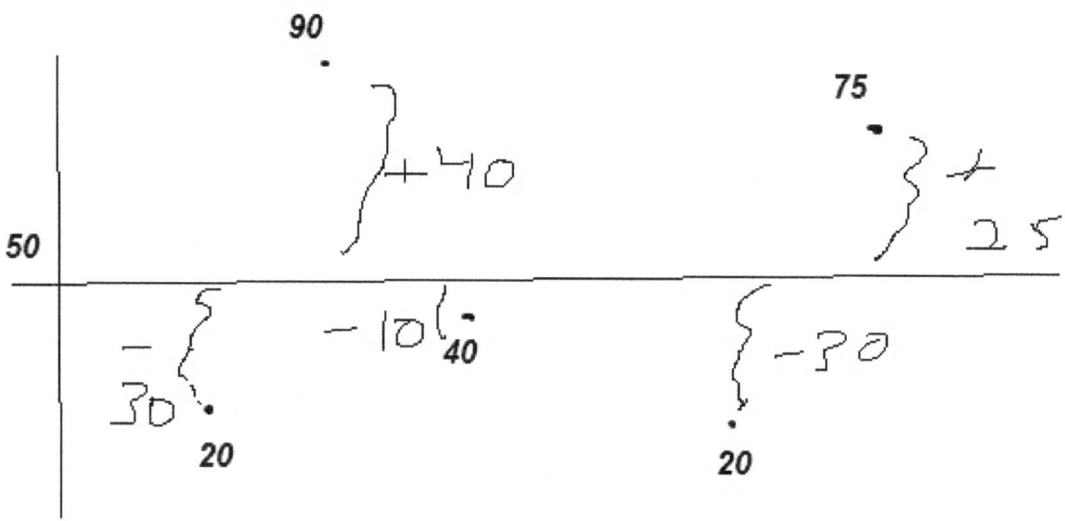

In the diagram above you can see the amounts below or above the average for each point. If the average actually is 50, the negative amounts must equal the positive amounts. Here the positive amount is 65 and the negative amount is 70. The average must be below 50 but not by much. Choices D and E are out. Let's try the next lowest choice, 1249 (or 49).

118 Chapter 4. ACT®-style Questions with Explanations

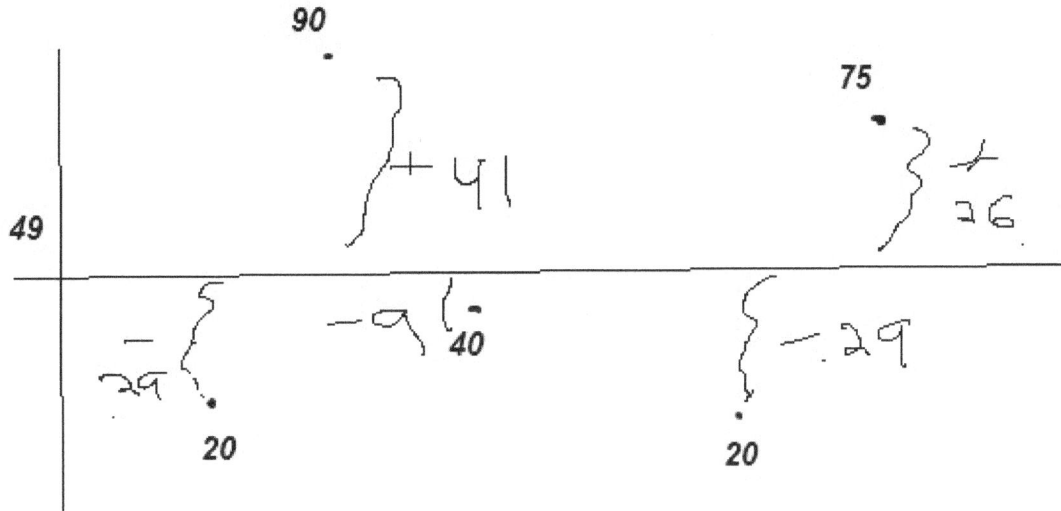

Now the positive numbers add up to 67. The negative numbers add up to 67. They balance each other. This proves that the average is 49 and choice C, 1249, is the correct answer.

---

44. Given the inequality $-5 < \frac{-5x}{2} < 15$, which statement below is equivalent?

F. $-10 < 5x < 15$
G. $-10 < 5x < 30$
H. $2 < x < 6$
J. $-2 < x < 6$
K. $-6 < x < 2$

Explanation of 44:

Remind yourself of what is tricky about inequalities. The test makers are definitely trying to see if you fall into a trap. In an inequality you can add or subtract the same amount from each side. You can multiply or divide each side by the same thing but **only** if the amount is positive.

If you multiply or divide two sides of an inequality by a negative number, the inequality signs reverse. As long as you are clear about that, you will not fall into the trap.

One good hint for a three-part inequality is that if you get confused, you can break the expression into its two separate components.

$$-5 < \frac{-5x}{2} \text{ and } \frac{-5x}{2} < 15$$

Glance at the answer choices. This is almost always a good strategy because it gives you an idea of what you are working with, what you have to distinguish, and what you do not have to distinguish.

The first two answer choices start with –10. The other three have some combination of 2's and 6's. The original inequality looks like it can easily be rewritten so that the first term becomes –10. Simply multiply each component by 2. The inequality signs stay the same because 2 is positive.

$$-10 < -5x < 30$$

Compare this with choices F and G. It is close but it doesn't match.

Consider choices H, J, and K. How can you manipulate the original expression to approximate one of those? Notice that all three have x as the middle term, without any numbers or fractions attached to it. That is a good clue. Other than that, there is no obvious way to get to 2 or 6, so you can go with the clue.

Simplify the inequality so that you can end up with x by itself in the middle. You have already multiplied all parts by 2, as shown above, so you can start with that. To get rid of the –5 in the middle you have to divide by –5. However, that means you need to reverse all of the inequality signs.

To maintain your accuracy, it might be helpful to divide all parts by positive 5 first.

$$-2 < -x < 6$$

Now we can see where the 2 and 6 come from! Divide by –1 in order to get x in the middle. Then you must reverse the inequality signs but again, it might help maintain your accuracy to do this in several steps.

    Step 1. Divide by –1.
    Step 2. Reverse inequality signs.
    Step 3. Switch the order of the inequality if necessary to match one of the answer choices.

$$2 < x < -6 \text{ (before reversing the inequality signs)}$$
$$2 > x > -6$$

This does not match choices H or J. To compare it with choice K, switch the order of the elements, being very careful to keep the inequality signs pointing in the correct direction. Start with –6. It is less than x.

$$-6 < x$$

Carefully double check that x is less than 2 and add this to the above.

$$-6 < x < 2$$

This matches choice K, which is the correct answer. It might feel like breaking the final steps down into these tiny bits is overkill but in reality many people will get this question wrong because they confuse the inequality signs in going from $2 > x > -6$ to $-6 < x < 2$. Part of being an effective test taker is knowing, and admitting, when you are likely to make a silly mistake and using powerful accuracy strategies to stay out of that trap.

45. For a matrix $\begin{bmatrix} m & n \\ o & p \end{bmatrix}$ the determinant is defined as $mp - no$. For the matrix $\begin{bmatrix} m & 6 \\ 5 & m \end{bmatrix}$, if $m < 0$, for what value of m is the determinant equal to m?

A. -30
B. -15
C. -6
D. -5
E. There is no negative value of m for which the determinant of the matrix equals m.

Explanation of 45:

Notice that you do **not** need to know what a matrix is or what the determinant of a matrix is. The test writers have thrown in the terms hoping to scare you away.

All you have to do is follow the simple directions they have given. For the matrix they give, the determinant would be $m*m - 6*5$ and the questions asks for what value of $m<0$ would that expression equal m.

$$m*m - 6*5 = m$$

You may notice that this is a quadratic equation. It may not be clear right now how to solve for m, but you can manipulate this as you would any quadratic equation to see if it is possible to find the factors of the equation.

But... before you do that, consider other methods for solving the problem. Examine the answer choices. We know that m is negative and all the choices are negative. What about testing the answer choices? Some of them are quite large and squaring them would get a little cumbersome but could be doable.

Try testing the smallest answer choice, choice D, -5.

$$-5 * -5 - 6 * 5 =$$
$$25 - 30 = -5$$

This works. You are done. Choice D is correct. If −5 had not been the correct answer, you would have had to square −6, -15, and −30.

$$36 - 30 =$$
$$225 - 30 =$$
$$900 - 30 =$$

You might have seen right away that these would all result in positive numbers and so could not be the correct answer.

As an exercise, try factoring the original equation, after bringing all of the elements over to the left side.

$$m^2 - m - 5*6 = 0$$

Leaving 5*6 as it is actually shows you the factors you need.

$$(m+5)(m-6) = 0$$
$$m = -5, 6$$

If you are comfortable with finding the factors of a quadratic equation, this would work but remember that testing answer choices might be faster and more accurate.

You finished Day 9! How did it go on these five questions?

Number of questions you got right on your own: _____

Types of problems or patterns you need more work on: _____

_____

How much new did you learn from these questions? ☐ Important tools! ☐ Some tools ☐ Not too much

Day 10, Questions 46-50

46. In a given right triangle with an angle α, sin α = 0.75. What is the value of cos α ?

F. $\dfrac{\sqrt{7}}{4}$

G. $\dfrac{\sqrt{7}}{5}$

H. $\dfrac{4}{5}$

J. $\dfrac{3}{5}$

K. $\dfrac{3}{\sqrt{7}}$

Explanation of 46:

Start by drawing the triangle. A visual image gives you much more intuitive information than trying to hold an image in your mind.

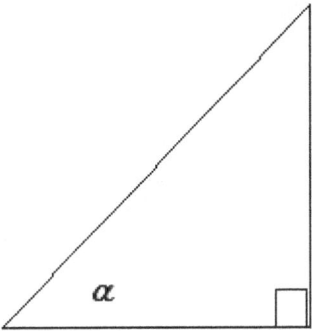

It is helpful to use this standard orientation, with the angle at the left. Because the sine is 0.75, or $\frac{3}{4}$, the side opposite the angle and the hypotenuse must have the ratio of 3:4.

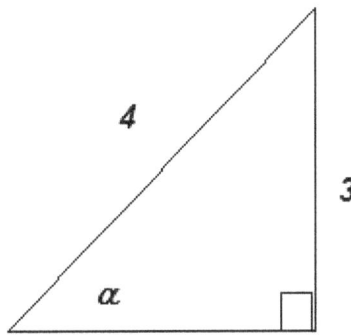

It does not matter whether the legs are actually 3 and 4 or some multiple of 3 and 4. The ratio is the same and the cosine is the same.

To calculate the cosine, you need to know the length of the base. Is this a 3:4:5 triangle?

Hopefully you noticed that it is not. A 3:4:5 right triangle must have the 3 and 4 as sides and the 5 as the hypotenuse. Remember that the hypotenuse is always the longest leg of the triangle.

Because this is not a 3:4:5 triangle, you have to use the Pythagorean Theorem to find the base.

$$3^2 + b^2 = 4^2$$
$$9 + b^2 = 16$$
$$b^2 = 7$$
$$b = \sqrt{7}$$

This means the cosine is $\frac{\sqrt{7}}{4}$. Choice F is the correct answer.

Review of Tan, Sin, Cos

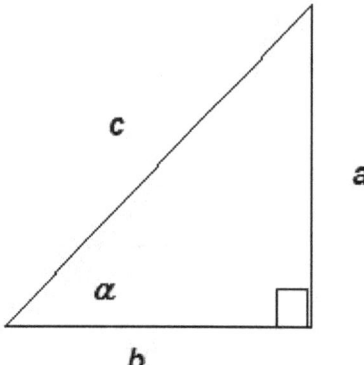

For angle α:

Tan α = $\dfrac{a}{b}$

Sin α = $\dfrac{a}{c}$

Cos α = $\dfrac{b}{c}$

If you have trouble remembering these, try to look for some associations that will help you. Tangent and cotangent are the only ones that do not use the hypotenuse.

Imagine yourself standing at alpha and look forward. What you see in front of you is *a*. What you are standing on is *b*.

---

47. Julio's scores on the previous five exams in History were 95, 83, 90, 76, and 86. After calculating his average for the five exams, Julio determined what he would have to score on the next exam to maintain the exact same average. What would he have to score?

A. 82
B. 83
C. 83.5
D. 85.2
E. 86

Explanation of 47:

The standard way to do this kind of problem is to add the five original scores and divide by 5. This is the average. In order to maintain the same average, his sixth score has to be exactly the average. The question, then, is really asking what the average of his first five scores is.

If you are confident in your skills with the calculator, you can use that where allowed. If you are not so confident, you can use intuitive methods to do the addition. For example, the method below lets you separate the tens from the ones and add them separately. This might allow you to be more accurate with a hand calculation.

95	90	5
83	80	3
90	90	0
76	70	6
86	80	6
totals	410	20

The five numbers add up to 430. Dividing that by 5 gives 86. The correct answer is choice E.

Another approach for this problem is to test the answer choices. The drawback is that you might have to test four of them. If you started with choice E, of course, you would have gotten the right answer immediately.

To test an answer, such as 86, you would use a diagram such as the one below, set up to show 86 as the average. Each point is placed above, below, or on the average line. You then mark how many units each point is above (positive) or below (negative) the average line. If the average is correct, the positives equal the negatives. In the figure below +13 balances out −13.

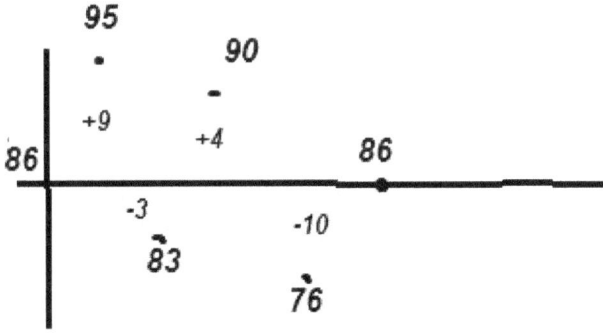

Chapter 4. ACT®-style Questions with Explanations

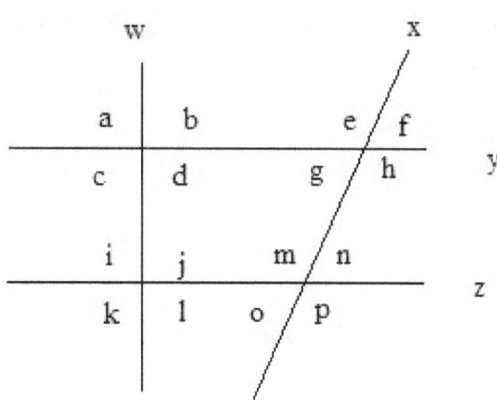

48. In the figure above there are four lines, w, x, y, and z and y∥z. The angles shown are *a* through *p*. If angle *e* measures 105°, which of the following is a complete list of the all the angles that must equal 75°?

F. h, m, p
G. f, g, m, p
H. f, g, n, o
J. f, g, n, o, b, c, j, k
K. h, m, p, a, d, i, l

Explanation of 48:

The key to this question is that *y* and *z* are parallel, so line *x* crosses *z* at the same angle that it crosses *y*. Does this make sense to you intuitively?

It is important to note that line *w* is not necessarily parallel to *x*. It might be, because you cannot assume figures are drawn to scale, but it clearly looks like it is not parallel. In any case you have to assume that it is not parallel unless the problem tells you that it is.

Because *w* cannot be assumed to be parallel to *x*, you know nothing about the angles that *w* forms with *y* and *z*. None of those angles must be equal to 75°. Scan the answer choices first for *a, b, c,* and *d*, because those angles cannot be in the correct answer. Choices J and K are out. Next scan for answer choices that contain *i, j, k,* and *l*. None of the remaining answer choices contain those.

The reason for breaking your scanning into two distinct searches is because it would be hard to scan for all eight angles of line *w* at the same time.

There are three answer choices left. Here is an important concept for working with angles formed by lines crossing. Unless the lines cross at a right angle, each of the four angles created by the crossing is either a big angle or a small angle. Each big angle is adjacent to a small angle. You can check this out in the diagram for the problem.

You can also notice that each big angle is across from another big angle. The same is true for the small angles. Small angles must be less than 90°. Big angles are greater than 90°.

Angle *e* is 105. It is a big angle. The angle across from it, *h*, is also a big angle. What about the angles adjacent to *e* – angles *f* and *g*? You may remember that angle *e* and angle *f* together form a straight line, so they have to add up to 180°. Subtracting 105 from 180 leaves 75°. Angles *f* and *g* are each 75.

Check the answer choices to see if you have eliminated anything. Choice A is out because it does not contain *f* or *g*.

A powerful strategy at this point is to examine the difference between the remaining two answer choices. This tells you what to focus on. One contains *m,p* and the other contains *n,o*. Both pairs are across from each other. Which one is 75? The answer is easy. It has to be the small angles, because only small angles are less than 90°.

The small angles are *n,o* and choice H is correct.

---

49. Katrina decides to make potholders to sell for a fund raising event. She starts on Sunday and continues through the following Saturday. She makes 50 potholders each day. Andy also makes potholders. He starts on Tuesday of the same week and continues through Saturday. He makes 70 potholders each day. By the end of Saturday how many potholders have the two made in total?

A. 350
B. 360
C. 420
D. 700
E. 710

Explanation of 49:

You do not need algebra for this. Use an intuitive way to organize the information for yourself so you do not get confused.

	Sun	Mon	Tues	Weds	Th	Fri	Sat	Totals
K	50	50	50	50	50	50	50	350
A			70	70	70	70	70	350

The total is 700. Choice D is correct. The problem is very easy once you have it organized intuitively. People who miss this question probably got caught up in trying to create an algebraic equation or got confused in the wording. A picture is worth at least a thousand words.

It is interesting to notice that 50 times 7 is the same as 70 times 5. The question could have asked you to figure out who made more potholders, when in fact they both made the same number.

Chapter 4. ACT®-style Questions with Explanations

50. For the function $f(x) = -5|x|^2$, what is $f(-5)$?

F. -625
G. 25
H. -25
J. -125
K. 125

Explanation of 50:

Orient yourself to what is going to be confusing about this question. Notice where the traps will lie. It has to do with negatives, doesn't it? Absolute values can confuse people between negatives and positives and the −5 in the expression can cause people to make errors with negatives and positives. If you know where the traps are, you are prepared for them!

Absolute Value Review

Absolute value can be thought of as the distance that a number is from zero on the number line. It is **always** positive.

$|3|$ and $|-3|$ are both equal to 3.

To find $f(-5)$ you plug the number 5 into the definition of $f(x)$.

$$f(-5) = -5(|-5|)^2$$

Before getting too involved in calculations, step back to see if there are any obvious clues here. Can you determine whether the answer is negative or positive?

The square of the absolute value of −5 must be a positive number. All squares are positive because negative numbers squared also come out positive. It turns out that it doesn't matter whether you take the absolute value of −5 or simply square −5. In both cases it comes out positive.

When you multiply a positive number times the first −5 in the expression, the answer is negative. Does that help eliminate any answers? Yes. Choice G and K are out. It is a powerful strategy to figure out now whether the answer is negative or positive because if you make a mistake in your calculations later, you will catch it.

Take a look at the remaining answer choices. They are quite far apart. This should make it easier to find the right answer because you do not have to worry about small differences.

Continue with the calculation.

$$-5(|-5|)^2$$
$$-5(5)^2$$
$$-5 * 25$$
$$-125$$

If you had gotten the sign wrong in your calculations, you would know you made a mistake because you already determined that the answer must be negative and you have crossed out all the positive answer choices.

Choice J is correct.

You finished Day 10! How did it go on these five questions?

Number of questions you got right on your own: _____

Types of problems or patterns you need more work on: _____

_____

How much new did you learn from these questions? ☐ Important tools! ☐ Some tools ☐ Not too much

Day 11, Questions 51-55

51. Given the equation $x^2 - 4x - 21 = 0$, the solution set of which of the equations below is the same as the solution for x?

A. $|x+6| = 3$
B. $|x+8| = 5$
C. $|x-4| = 3$
D. $|x+2| = 5$
E. $|x-2| = 5$

Explanation of 51:

The question asks you to find an equation for which x has the same solution set as the original quadratic equation. In some questions like this, you have to manipulate the original equation into an equivalent form. However, that is not the case in this question. The equations in the answer choices are not derived from the original equation.

The original equation is quadratic and has two solutions for x. The equations in the answer choices also have two solutions for x but it is because of the absolute value. If the absolute value of x is 3, x could be 3 or -3.

What is your plan of attack? You first have to find the two solutions for the original equation and then evaluate the answer choices to see which one has the same solutions. There are some short cuts you can use when you get to that step.

What if you cannot find the two solutions to the original equation? No problem. You can start with the answer choices, finding the two solutions for each one and then plugging those into the original to see if they work. For the correct answer choice, they will.

There are several steps here. Create a road map to maintain your accuracy.

> Step 1. Determine the two solutions for x in the original equation.
> Step 2. Find the two solutions in each answer choice. Look for short cuts.
> Step 3. Find the answer choice that matches the solutions to the original.

Step 1. To find the two solutions to a quadratic equation, rewrite the equation as the product of two sums.

$$x^2 - 4x - 21 = 0$$
$$(x + \ )(x + \ ) = 0$$

The two numbers that go in the blanks have to have a product of negative 21. Then, when you foil the expression, you have to end up with $-4x$.

If you are a little shaky on this process, let's break it down a little. Let's call the missing numbers a and b.

$$(x + a)(x + b) = 0$$

When you foil this, you get:

$$x*x + a*b + ax + bx$$
$$x^2 + ax + bx + ab$$
$$x^2 + (a + b)x + ab$$

If you are not too clear on how we got to each step, take a minute to look it over.

Now you can see the restrictions on a and b.

$$ab = -21$$
$$a + b = -4$$

Sometimes it is possible to spot the correct numbers at a glance. Other times you need an intuitive tool. A good place to start is with the numbers that multiply to give -21. Let's organize the options and test whether they add up to –4.

The positive number 21 can be broken into two factors in only two ways – 1 times 21 and 3 times 7. For negative 21, we also have to consider the signs of these factors.

a,b	a + b
1, -21	-20
-1, 21	20
3, -7	-4
-3, 7	4

Because you know that one of the traps of this question is negative versus positive, for each addition first ask yourself which number is "larger" meaning further from zero on the number line. That number determines the sign of the answer. For example, in the first row, the –21 has the larger absolute value so the answer will be negative.

Only the third row in the table gives –4, so the correct numbers for a and b are 3 and –7. Which one is a and which one is b? It does not matter. Here is how it comes out.

$$(x + 3)(x + -7) = 0$$

There is one more trap in your path. You are looking for the solutions for x. At this point, you only have the two factors for the equation. Let's break down the process for finding x itself. If you already feel confident on this, this will be a good review.

If $y*z=0$, then either $y=0$ or $z=0$. Is that clear? The only way to multiply some numbers and end up with zero is if one of them is zero.

$$(x + 3)(x + -7) = 0$$
$$(x + 3) = 0$$
$$\text{or}$$
$$(x + -7) = 0$$

Now solve each equation for x.

$$x = -3 \text{ or } x = 7$$

It is very important to be absolutely sure of the signs of your answers. We now have the two values of x that you have to match in the answer choices. This completes step 1.

Step 2. Instead of evaluating each answer choice, there is a short cut. Remember that the correct answer has to work for both values of x. Let's first test for 7. Any answer choice for which 7 does not work is out. That will limit what else you have to test. We choose 7 because a positive number is more accurate to work with.

Test choice A. 7 + 6 is clearly not 3. Test choice B. 7 + 8 is not 5. Test choice C. 7-4 is 3. Leave choice C in. Test choice D. 7 + 2 is not 5. Test choice E. 7 – 2 is 5. Only choices C and E remain. This is a lot easier than having to test five answer choices.

Next we can test –3 in choices C and E, being very careful to be accurate with signs. In choice C, $|x-4| = 7$, not 3. Choice C is out.

As a double check, test choice E. $|-3+-2| = |-5| = 5$. Choice E is confirmed as the correct answer.

---

52. The two equations below define the variables x and y. Which of the following accurately represents all possible values for x and y?

$$5x - 3 = 8$$
$$y^2 - 7 = 9$$

F. $x = \dfrac{11}{5}$ and $y = 4$

G. $x = 1$ and $y = 4$ or -4

H. $x = 2\dfrac{1}{5}$ and $|y| = 4$

J. $x = -1$ and $y = 4$ or -4

K. $x + y = 6\dfrac{1}{5}$

Explanation of 52:

You might be tempted to treat this problem like a set of equations in x and y that have to be added or subtracted or set equal to each other or some other algebraic manipulation. This is not necessary.

Notice that the first equation only involves x. You can simply solve for x. The second equation only involves y. You can solve for y. The two equations are not related to each other in any way.

You can find x and then scan the answer choices to eliminate any that do not match x. After that, evaluate the remaining choices for y.

$$5x - 3 = 8$$
$$5x - 3 + 3 = 8 + 3$$
$$5x = 11$$
$$x = \dfrac{11}{5}$$

Which answer choice can be eliminated? Choices G and J are clearly out. What about choice H? $\dfrac{11}{5}$ is actually equivalent to $2\dfrac{1}{5}$. It stays in. Choice K is difficult to evaluate. You have to leave it in until you have a value for y.

$$y^2 - 7 = 9$$

$$y^2 - 7 + 7 = 9 + 7$$
$$y^2 = 16$$
$$y = 4, -4$$

The trick here is to remember that there are two solutions for square root of 16, a positive one and a negative one.

Test the remaining answer choices. Choice F is incorrect because it omits –4. Choice H appears to be correct because it does include both 4 and –4. Test choice K to make sure you have not made a mistake.

Choice K works for the positive value of y but not for the negative value. Choice H is confirmed as the correct answer.

53. Four people – Monty, Nan, Oliver, and Pedro – stand on the perimeter of a rectangular garden plot with a length of 20 feet and an area of 240 square feet. Monty stands at the middle point of one of the long sides. Nan stands exactly 12 feet to the left of Monty. Oliver stands exactly 18 feet to the right of Nan. Pedro stands exactly 5 feet to the left of Oliver. Starting with Monty and going to the right around the perimeter of the rectangle, what is the order of the four people?

A. M, O, P, N
B. M, N, O, P
C. M, P, O, N
D. M, P, N, O
E. M, N, P, O

Explanation of 53:

This is the perfect problem for a diagram. The test makers have made this problem look more complex than it is. Let's start with some basics and then see if it is necessary to go into more detail. That way we can avoid doing more work than necessary.

Draw the rectangle. You can label one side 20 but it might not be clear yet whether 20 is a long side or short side. For now, assume that it does not matter.

Place Monty in the middle of the top long side. Follow the directions, person by person. Nan is 12 feet to the left of Monty. That is easy to place in the diagram.

Oliver is 18 feet to the right of Nan. That takes us back to Monty and then 6 feet beyond. Finally, Pedro is 5 feet to the left of Oliver, which takes us almost back to Monty but not quite.

Chapter 4. ACT®-style Questions with Explanations

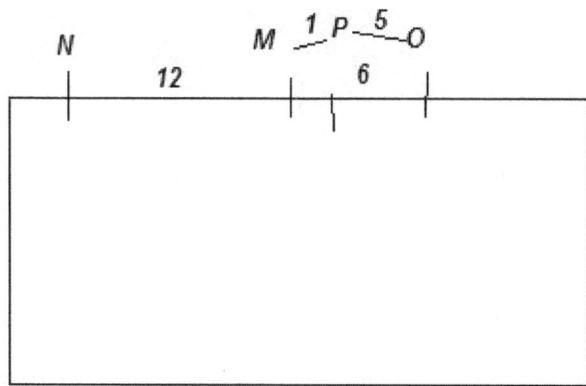

The question only asks for the order of the people. You have that. There is no need to calculate the length of the second side. It does not matter if N, for example, ends up wrapping around to the left side. Only the relative order matters and you have already determined that. Starting with Monty and going around to the right, the order is M, P, O, N. Choice C is correct.

54. If $f(x) = \sqrt{x} - 3$, and if $y = x^2 - 12x + 36$, which of the following is equivalent to $f(y)$?

F. $x - 6\sqrt{x} + 6$
G. $x + 3$
H. $x - 9$
J. $x - 6$
K. $x^2 - 12x + 33$

Explanation of 54:

Notice what is complicated about this problem. It will be challenging to write the correct expression for $f(y)$. To be accurate, go step by step and write each step down on your scratch paper. To avoid errors, make only one small change on each step.

$$f(y) = \sqrt{y} - 3$$
$$= \sqrt{x^2 - 12x + 36} - 3$$

In the second line we have replaced y with the definition of y.

It would be easy to simplify this expression if it turned out that $x^2 - 12x + 36$ was a square. If not, it would be extremely difficult to simplify the expression. You can assume that the test will not make this impossibly hard. See if you can divide the expression into two equal factors.

$$x^2 - 12x + 36$$
$$(x + a)(x + a)$$

To end up with +36, the missing term, a, must be either 6 or −6. To end up with −12x, only the −6 will work.

$$(x + -6)(x + -6)$$

Now we can go back to the expression for $f(y)$.

$$\sqrt{x^2 - 12x + 36} - 3$$
$$\sqrt{(x-6)(x-6)} - 3$$
$$(x-6) - 3$$
$$x - 9$$

Choice H is the correct answer.

55. When strong winds were predicted in her area, Alicia began tracking wind speeds in 30-minute intervals, i, beginning with interval 0. Her results in miles per hour are shown in the table below.

Interval, i	Speed, mph, s
0	3
1	10
2	17
3	24
4	31
5	38

Which of the following expressions accurately represents the relationship between i and s, as shown in the table?

A. $s = 7i + 3$
B. $s = i + 3$
C. $s = 3i + 7$
D. $s = 10i + 3$
E. $s = 5i + 7$

Explanation of 55:

There may be some complex mathematical way to find the relationship from the data but it is not necessary. Simply take one set of data and see which answer choice or choices match it.

Start with $i = 0$ and $s = 3$.

Choice A = 3. That fits.
Choice B = 3. That fits.
Choice C = 7. It is out.
Choice D = 3. That fits.
Choice E = 7. It is out.

Do the same with the second data point - $i = 1$ and $s = 10$.

Choice A = 10. It is in.
Choice B = 4. It is out.
Choice D = 13. It is out.

Choice A is the correct answer.

You finished Day 11! How did it go on these five questions?

Number of questions you got right on your own: _____

Types of problems or patterns you need more work on: _____

_____

How much new did you learn from these questions? ☐Important tools! ☐Some tools ☐Not too much

Day 12, Questions 55-60

56. Given the equation $y^2 - 10y = 24$, what are the solutions for y?

F. 4, -6
G. -4, 6
H. 2, -12
J. -2, 12
K. 0, -24

Explanation of 56:

This question asks you to find the solutions for y in a quadratic equation. You can assume that the test will give you equations that can be factored easily. Here is a review of the process.

$$y^2 - 10y = 24$$
$$y^2 - 10y - 24 = 0$$
$$(y + a)(y + b) = 0$$
$$ab = -24$$
$$a + b = -10$$

Assume that the test will only give you quadratic equations for which the a and b are whole numbers. Find the pairs of numbers that give the product 24. You will deal with whether they are positive or negative later. The pairs are 1 and 24, 2 and 12, 3 and 8, 4 and 6.

Let's organize this information in a table. Because *ab* must be negative, only use pairs with one negative number

a,b	a + b
-1,24	23
1,-24	-23
-2,12	10
2,-12	-10

You can stop there because you have found the pair that produces –10 for *a + b*.

$$(y + 2)(y + -12) = 0$$

The pitfall at this point is to mistakenly believe that 2, -12 are the values of *y*. Remember that if *wz*=0, then either *w* or *z* must equal zero.

$$(y + 2) = 0, \text{ or}$$
$$(y + -12) = 0$$

So,

$$y = -2, \text{ or}$$
$$y = 12$$

Choice J matches these values and is the correct answer.

---

57. Given $f(x) = \dfrac{x}{x+1}$ and $g(x) = \dfrac{x+1}{x}$, what is the value of $f(g(2))$?

A. $\dfrac{2}{3}$

B. $\dfrac{3}{2}$

C. $\dfrac{3}{5}$

D. $\dfrac{5}{3}$

E. $\dfrac{2}{3} * \dfrac{1}{5}$

Explanation of 57:

Take this step by step. First, find the value of *g*(2). Then plug that number into the expression for *f*(*x*). Try writing down this plan as a road map.

Chapter 4. ACT®-style Questions with Explanations

To find g(2), plug 2 into the definition of g(x).

$$g(x) = \frac{x+1}{x}$$

$$g(2) = \frac{2+1}{2}$$

$$= \frac{3}{2}$$

Let's leave $\frac{3}{2}$ in fractional form for now. Plug it into the definition of f(x).

$$f(x) = \frac{x}{x+1}$$

$$f(\frac{3}{2}) = \frac{\frac{3}{2}}{\frac{3}{2}+1}$$

You will probably have to simplify that fraction. First, though, look at the answer choices to see what we are dealing with. One powerful strategy with estimating fractions is to determine whether the fraction is more than or less than 1. The numerator is smaller than the denominator in the fraction above, so it is less than 1. Eliminate any answer choice that is larger than 1. Choices B and D are out.

Compare the remaining answer choices to see what you are dealing with. Choices A and C are relatively close to each other. Choice E is much smaller.

Because choices A and C are close, you must continue with simplifying the fraction for $f(\frac{3}{2})$. You have two options. Simplify the fraction using the fractions in the numerator and denominator or convert those fractions into decimals. Let's try the decimal approach first.

$$\frac{1.5}{1.5+1}$$

$$\frac{1.5}{2.5}$$

$$\frac{15}{25}$$

$$\frac{3}{5}$$

That is pretty straightforward and shows that answer choice C is correct.

Let's see what would happen if you work with fractions instead of decimals.

$$\frac{\frac{3}{2}}{\frac{3}{2}+\frac{2}{2}}$$

$$\frac{\frac{3}{2}}{\frac{5}{2}}$$

$$\frac{3}{5}$$

58. For two real nonzero numbers *m* and *n*, which of the following expressions must *always* be less than zero?

F. $m^2 - n^2$
G. $m - n$
H. $\sqrt{m^2} - \sqrt{n^2}$
J. $|m| - |n|$
K. $-|m| - |n|$

Explanation of 58:

Orient yourself carefully to the question stem before doing anything else. What are real numbers? Pretty much anything. Only imaginary numbers do not belong to the set of real numbers. The only restriction on *m* and *n*, then, is that they cannot be zero.

What is the question asking? You have to find the answer choice that must always be negative. One way to solve this is to test each answer choice. To do this, you try to eliminate it by finding an example in which the expression is not less than zero. In other words, any result that is positive or zero proves that that answer choice does not have to be less than zero.

Finding an example that makes the answer choice zero is often the easiest way to go. Start with choice F. Make *m* and *n* the same and choice F will equal zero. Choice F is out.

Choice G. The same strategy works here. Choice G is out.

Choice H. Same thing. Choice H is out.

Choice J. Same thing. Choice J is out.

Chapter 4. ACT®-style Questions with Explanations

Choice K. $-|m|$ must be a negative number. $|n|$ is positive. If you start with a negative number and subtract a positive amount, the answer must be negative. Choice K is the correct answer.

If you wanted to find an exception that was positive for each answer choice, you could just make *m* larger than *n*. To keep it concrete, you can use simply numbers like 1 and 2.

---

59. $\frac{1}{10}$ of what number is equivalent to 11% of 40?

A. 440
B. 44
C. 0.44
D. 0.40
E. 0.11

Explanation of 59:

This problem does involve some algebra but keep it simple by going one step at a time. First calculate what 11% of 40 is. You can use an intuitive tool. Find 10% of 40 and 1% of 40 and add them together.

100%	40
10%	4
1%	0.4
10% + 1% = 11%	4 + 0.4 = 4.4

The problem now is:

$$10\% \text{ of } ? = 4.4$$

One simple way to approach this is to test the answer choices. It is much easier to determine what 10% of a number is than to determine what number you would have to have so that 10% of it is 4.4.

10% of Choice A. 44
10% of Choice B. 4.4
10% of Choice C. 0.044
10% of Choice D. 0.04
10% of Choice E. 0.011

Choice B is the correct answer.

60. A solid object placed in water displaces a volume of water equivalent to its own volume. This method can be used to determine the volume of an irregularly shaped object. A physics student pours water into a rectangular tub measuring 25 inches by 30 inches and fills the tub to a depth of 8 inches. The student then submerges a crown in the water. The water level rises to 12 inches. What is the volume of the crown in cubic inches?

F. 9000
G. 6000
H. 3000
J. 1200
K. 1000

Explanation of 60:

This is a bit of a physics problem. If you orient to it carefully, you might determine that you need to find the change in volume and the change is the final volume minus the original volume. That is the road map.

    Step 1. Determine the original volume.
    Step 2. Determine the final volume.
    Step 3. The change is step 2 – step 1.

Step 1. The original dimensions of the water are 25x30x8. Check to make sure all dimensions are in the same units (inches.) The volume is the product of the three dimensions.

To do the multiplication intuitively, break the numbers up into factors that are easier for you to work with.

$$25 * 4 * 2 * 30$$
$$(25 * 4)(2 * 30)$$
$$100 * 60$$
$$6000$$

Step 2. The final dimensions are 25x30x12.

$$25 * 4 * 3 * 30$$
$$(25 * 4)(3 * 30)$$
$$100 * 90$$
$$9000$$

Step 3. 9000 – 6000 = 3000. Choice H is correct.

You might have noticed that the difference between the original volume and the final volume was really just 4 inches of the same 25x30 tub. You could get right to the increase in volume by calculating the volume of 4 inches.

$$25 * 4 * 30$$

Chapter 4. ACT®-style Questions with Explanations

$$100 * 30$$
$$3000$$

You finished Day 12! How did it go on these five questions?

Number of questions you got right on your own: _____

Types of problems or patterns you need more work on: _____
_____

How much new did you learn from these questions? ☐ Important tools! ☐ Some tools ☐ Not too much

Congratulations on finishing Chapter 4

Chapter 5. SAT®-style Non-Calculator Questions with Explanations

The questions in this chapter are written and formatted in the style of questions on the SAT® exam. They represent the most common patterns of math questions found on both the ACT® and SAT® exams. You are not allowed to use a calculator on this section. There is a second SAT® math section that does allow you to use a calculator.

Try to work each problem on your own before reading the explanation. Take as much time as you need. It can be helpful to give yourself 15 or even 30 minutes or more to work on a problem. The longer you work on it, the more you can learn.

The explanations are designed to be simple and intuitive. Nevertheless, you will probably find some explanations challenging. Stick with it. Experiment with it. Work on it with a friend.

In some problems you might find that the intuitive strategies seem unnecessary and that the problem can be easily solved with standard math tools. We suggest that you still study the intuitive tools for that problem. You may need those tools on another problem.

Your primary focus is to learn new ways of thinking about mathematical relationships. It does not matter so much whether you get a question right or wrong now. The patterns in this chapter are the patterns you will see on your test. Study them, learn them, and get comfortable with intuitive tools for solving them.

You can follow the daily assignments – about five questions per day – or you can do more or fewer questions per day. At the end of each day's assignment you can evaluate how you did.

Day 13, Questions 1-5

1. Shannon types an average of 40 words per minute and Nick types an average of 45 words per minute. If Shannon types for s minutes and Nick types for n minutes, which of the following expressions represents the total average number of words typed by Shannon and Nick?

A) $85(s + n)$

B) $42.5(s + n)$

C) $45s + 40n$

D) $40s + 45n$

Explanation of 1:

This question tests your understanding of rates. A rate is a ratio between a quantity (in this case, words) and a time period (minutes.)

40 words per minute is a good example of a rate. It consists of two parts – the quantity of the product (40 words) and the time period (1 minute).

Don't be fooled by the word "average" in the problem. This problem is not about averages. The test uses the word average because people don't type at exactly the same rate every minute. They might type faster for a while and then slow down. By citing an average rate, the test wants you to treat the rate as though each person typed exactly at the given rate all the time.

In this problem there are two different rates - one for Shannon and one for Nick. The question asks you to find the expression that gives the total number of words typed by the two. It might seem tempting just to add the two rates together or to take an average of the two rates. There **is** a way that you can do that to get to the right answer but it is fairly complex and you are just as likely to get confused.

What is the simplest way to figure out the total product (words) of two people with two different rates for two different amounts of time? Just calculate how many words Shannon typed, how many words Nick typed and add the two together.

In case you're not too clear on how to figure out how many words each person typed, let's take a look at how you can factor the units (words, minutes) associated with each number to make sure you have your calculation set up correctly.

$$\frac{40 \text{ words}}{1 \text{ Minute}} * \frac{s \text{ Minutes}}{1} = \frac{40s \text{ Words}}{1}$$

Notice that the units "minutes" cancel out. There is one on top of a fraction and one on the bottom. That means the final unit is simply "words", which is what you are looking for.

I call this "factoring units." It's a powerful way to make sure you have everything set up right.

Using this method, the Shannon types 40s words and Nick types 45n minutes. Add those two numbers together and you have answer choice D.

There is a different way to solve this problem that uses a powerful strategy when there are variables in the answer choices. In this problem the answer choices contain the variables s and n.

In this strategy you simply assign numbers to the variables. Watch how this works here. Let's say that Shannon times for 5 minutes ($s = 5$) and Nick types for 10 minutes ($n = 10$). Now you can easily calculate that Shannon types 200 words (40 words/minute times 5 minutes) and Nick types 450 words (45 words/minute times 10 minutes.) The total is 650 words.

Now use these same values for s and n in the answer choices and see which one comes out to 650. That one will have to be the correct answer.

There are two important cautions with this method. More than one answer choice might work. In that case, any answer that didn't work is out, but you will need to choose different numbers for your variables until you are down to one answer.

The second caution is that you have to follow any restrictions that are placed on the variables. For example, if the problem had said that Shannon typed for twice as long as Nick, you would have to choose numbers that meet that condition.

$$(3a^2 + 25ab^2 - 9b^2) - (-2ab^2 - 9b^2 + 4a^2)$$

2. The expression above is equivalent to which of the following?

A) $7a^2 + 23ab^2 - 18b^2$

B) $ab^2 + 16b^2 - 5a^2$

C) $-a^2 + 27ab^2$

D) $34a^2b$

Explanation of 2:

This is a very abstract question because it is just pure algebra – no real world context. That means that it will be easy to get lost and hard to stay accurate. Take a minute to identify some things that will make this problem confusing.

One of the most confusing elements is that there are a lot of subtractions and a lot of negative numbers. Adding positive numbers is relatively easy to keep track of. Adding negatives is more prone to error and subtracting negatives is even more so.

Fortunately, there is an easy way to turn negatives into positives and subtractions into additions. Study this simple example:

Original:

$$7 - 3$$

Rewrite as:

$$7 + -3$$

Is it clear that those are equivalent and that the second calculation avoids having to use subtraction. (Yes, it's true that you end up subtracting 3 from 7, but when you have a complex string of calculations, using addition will be more accurate.)

Take a closer look at how we converted the first expression into the second.

$$7 - 3$$

$$7 + (-1 * 3)$$

$$7 + -3$$

Does that make sense? So, if you have something like:

$$(a) - (b),$$

you can rewrite it as:

$$(a) + -1(b)$$

With a complex expression, you made need to then distribute the −1 over all of the elements in b:

$$(10 + 5) - (3 + 7)$$
$$(10 + 5) + -1(3+7)$$
$$(10 + 5) + -3 + -7$$

Another advantage of turning subtractions into additions is that additions can be done in any order, whereas subtractions can't. In the expression above, we can reorder the four elements in any way.

$$10 + -3 + -7 + 5$$

Coming back to the original problem, a good first step is to rewrite it by changing subtractions to additions.

$$(3a^2 + 25ab^2 - 9b^2) - (-2ab^2 - 9b^2 + 4a^2)$$

$$(3a^2 + 25ab^2 + -9b^2) + -1(-2ab^2 + -9b^2 + 4a^2)$$

Now you can distribute the −1 over the second expression in parenthesis:

$$(3a^2 + 25ab^2 + -9b^2) + (2ab^2 + 9b^2 + -4a^2)$$

Now that everything is addition, you can rearrange terms to combine like ones. Notice that there are three types of terms - a^2, ab^2, and b^2. For each, there are two terms. When you rearrange the expression to put the like terms together, do you get something like this?

$$3a^2 + -4a^2 + 25ab^2 + 2ab^2 + 9b^2 + -9b^2$$

If you combine the like terms, you should get the answer:

$$-a^2 + 27ab^2$$

Notice that the b^2 terms cancel out.

An alternative approach to this question would be to assign values to the variables. Notice that there are variables in the answer choices and that is your signal that you might try assigning values. Because the numbers 1 and 0 have special properties, you should probably not use them. You could make *a* equal to 2 and *b* equal to 3. Try that on your own and see if that makes it easy to calculate an actual value for the original statement and then find if there is one answer choice with the same value.

$$f(x) = c\sqrt{x} + 3$$

3. In the function *f* given above, *c* is a constant. $f(36) = 15$. What is the value of $f(100)$?

A) 203

B) 23

C) $3\sqrt{10} + 3$

D) $2\sqrt{10} + 3$

Explanation of 3:

Don't be intimidated by *f(x)*. This is a *function*, which simply means that it is defining a process that it wants you to follow. Here's a simple, understandable example.

The function of *x* is to bounce *x* three times and then pass it to Cherie.

If *x* is a volleyball, you bounce it three times and pass it to Cherie. If *x* is a basketball, you bounce it three times and pass it to Cherie.

It's that simple. Now let's take a closer look at the process that this problem defines. Take *x*, find it's square root, multiply that square root by the constant *c*, and then add 3 to the result. Not too complicated, right?

The problem doesn't come out and tell you what *c* is but because they tell you what *f(x)* is for a specific *x*, you can figure out *c* and then use that to figure out *f*(100).

If you are still feeling a little intimidated because of all of the variables in the expression $f(x) = c\sqrt{x} + 3$

try making yourself more comfortable by plugging in actual numbers. That usually makes expressions easier to understand. Use the numbers that the problem provides.

$$f(36) = c \text{ [square root 36]} + 3 = 15$$

$$c(6) + 3 = 15$$

If you didn't know the square root of 36, that's ok but for the exam you should memorize squares of at least 2 through 10 and maybe even through 15.

Now you can use simple algebra to figure out what c is. However, there are also some intuitive strategies that you might feel more comfortable with. Consider:

$$[\text{black box}] + 3 = 15$$

What would have to be in the box for this equation to work? Maybe your intuitions tell you that the box has to be 12. This is an intuitive insight. You don't have to do any math to get that insight.

Now the box is the same as $c(6)$, right? So you have figured out that

$$c(6) = 12$$

Using the black box method, what does c have to be?

$$[\text{new black box}] \times 6 = 12$$

Does your intuition tell you it has to be 2? If so, then you have discovered that c is 2 without having to do the algebra or even any additions or subtractions.

Now that you know that c is 2, you can solve for $f(100)$.

$$f(100) = c \text{ [square root 100]} + 3$$
$$f(100) = 2 \text{ [10]} + 3$$
$$f(100) = 20 + 3 = 23$$

---

4. A bag of 25 apples and a bag of 25 oranges are left out and forgotten in the back of a storeroom on August 1. The fruits will start to dry up, losing a certain amount of weight each week. If, on a given week, the weight of the apples is expressed in pounds as $15 - 0.6w$ and the weight of the oranges is expressed as $12 - 0.3w$, where w is the number of weeks that have elapsed since August 1, what will the weight of the bag of apples be when the two bags weigh the same amount?

A) 14.5 pounds

B) 13.5 pounds

C) 9 pounds

D) 3 pounds

Explanation of 4:

This is a fairly complex story problem but it has a lot of information that you can use to solve the problem. You simply have to have tools for organizing the information. The first step is to notice that we are comparing two things. The tests love to use contrasts between two categories. You have to be careful to keep the two separate and then you need to make notes of what you know about each category. The best way to do this is to use two columns on your scratch paper.

	Apples	Oranges
formula	$15 - 0.6w$	$12 - 0.3w$

Now, here is another very common pattern on the test. The two items are changing – decreasing in weight each week – at different rates. Do you see that the apples are losing more weight each week? Each week reduces their weight by 0.6, whereas the oranges are only reduced by 0.3. However, the apples started out at a higher number (15), as opposed to 12 for the oranges. (You can get these numbers by setting w to 0.) Because the apples start higher but lose weight faster, at some point they'll weigh less than the oranges.

You may be tempted to use create an algebraic equation to figure out at what number of weeks (w) the apples will weigh the same as the oranges. Of course, it is possible to do that but many people will have trouble doing it accurately. Let's look at a more intuitive way.

We can start with something simple that's easy to understand. Let's figure out what this scenario will look like at the end of one week. This is the "snapshot" approach. We are taking a snapshot of what the scenario will look like in one week.

	Apples	Oranges
formula	$15 - 0.6w$	$12 - 0.3w$
Week 1	$15 - 0.6(1) = 14.4$	$12 - 0.3(1) = 11.7$

The apples still weigh more, by a significant amount. Now, look at the answer choices and see if this new information helps you. At week 1 there are 14.4 pounds of apples. When apples equals oranges, there will have to be fewer than 14.4 pounds. Choice A is out. Choice B doesn't seem quite far enough away from 14.4 but it's hard to tell for sure. Choice D looks like it may be too far away.

Let's try another week. We should pick a number that's easy to calculate with, for example 10.

	Apples	Oranges
formula	$15 - 0.6w$	$12 - 0.3w$
Week 1	$15 - 0.6(1) = 14.4$	$12 - 0.3(1) = 11.7$
Week 10	$15 - 6 = 9$	$12 - 3 = 9$

We got lucky! At week 10, both weigh 9. If it had turned out that the apples weighed less than the oranges at week 10, you could have tried something between week 1 and week 10. It's usually best to start with numbers that are easy to calculate with, so you don't make a calculation error.

$$(sy + 8)(ty + 5)$$
$$5y^2 + ny + 40$$

5. If the two expressions above are equivalent for all values of y, and if $6 - t = s$, which answer choice below contains two numbers both of which could be a value for n?

A) –5, -8

B) 6, 13

C) 13, 33

D) 33, 45

Explanation of 5:

You can start organizing this by using the fact that the two expressions are equivalent (equal).

$$(sy + 8)(ty + 5) = 5y^2 + ny + 40$$

Notice that one side of the equation is a quadratic expression and the other side is a product. If we want to compare these two, we have to FOIL the left side.

$$sty^2 + 8ty + 5sy + 40 = 5y^2 + ny + 40$$

Both sides have a y^2 term. On the left side the y^2 is multiplied by st. On the right side it is multiplied by 5. This means that one possible solution is that st equals 5.

$$st = 5$$

Continue to simplify the equation to make the two sides more similar in structure. Both sides have 40, so the 40 can be subtracted from both.

$$sty^2 + 8ty + 5sy = 5y^2 + ny$$

See if it is possible to rewrite this in a way that the left side has a term that is something times y.

$$sty^2 + (?)y = 5y^2 + ny$$

In the expression $(8ty + 5sy)$, both elements do contain a y so you can "undistributed" the expression as

$$sty^2 + (8t + 5s)y = 5y^2 + ny$$

Can you compare these now? *St* can equal 5 and *(8t + 5s)* can equal *n*. In other words the *st* on the left side serves the same function as the 5 on the right. Likewise, the *(8t + 5s)* on the left serves the same function as the *n* on the right. You purposely rewrote the equations to make the two sides parallel in structure.

Now you know two concrete facts. $st = 5$ and $6 - t = s$. The second one tells you the relationship between *t* and *s* but not in a very intuitive way. Rewrite it as: $6 = s + t$. Now you can find numbers for *s* and *t* that satisfy both equations.

$$st = 5 \text{ and } s + t = 6$$

If you stick with the simplest numbers, positive integers, only 1 and 5 have the product 5. Fortunately, these two also add up to 6. The test tends to make these numbers easy to find.

The final step is to find what *n* equals when *t* and *s* are 1 and 5. There are two possibilities, depending on whether you assign 1 or 5 to *s*.

$$n = (8t + 5s)$$
$$n = (8*1 + 5*5) = 8 + 25 = 33$$
$$n = (8*5 + 5*1) = 40 + 5 = 45$$

Choice D is correct.

You finished Day 13! How did it go on these five questions?

Number of questions you got right on your own: _____

Types of problems or patterns you need more work on: _____

_____

How much new did you learn from these questions? ☐ Important tools! ☐ Some tools ☐ Not too much

Day 14, Questions 6-10

$$5x = 23$$

6. Given the above, if $10x + c = 50$, what is the value of *c*?

A) 4

B) 4.6

C) 23

D) 46

Explanation of 6:

This looks pretty algebra-y but there is a great intuitive shortcut, so you will **not** need to solve one equation for x or add equations.

Here's the trick. The first equation is in terms of $5x$. The second equation contains $10x$. Do you see the relationship? Try just doubling the first equation (multiplying **both** sides by 2).

$$10x = 46$$

Now this equation and the second are in very similar forms. Compare:

$$10x = 46$$
$$10x + c = 50$$

Do you see the intuitive answer? How much do you have to add to the right side of the first equation to get to 50? If you add the same amount to the left side, the two equations will be equal.

$$10x + 4 = 46 + 4$$

c is 4.

If you can't quite get that last step, once you have

$$10x = 46$$
$$10x + c = 50$$

, you can replace the $10x$ with 46 and rewrite as: $46 + c = 50$

---

7. A factory produces toys of various sizes. Each day the factory produces one size of toy. The number of toys that the factory produces in a day is expressed as $wCvh$, where w is the number of workers in the factory that day, C is a constant, v is the volume in cubic inches of the toy being produced, and h is the number of hours that the factory operates that day. If all of the potential workers attended a one-day training to help them be more productive, which of the factors in the expression would change on the next work day?

A) w

B) C

C) v

D) h

Explanation of 7:

To answer this question, you need to carefully understand what each variable represents and the relationship between them.

W is the number of workers. If there are more workers (i.e., if *w* increases) what happens to the number of toys produced?

C is a constant that is needed for the numbers to come out correctly.

V is the size of the toy. If the size of the toy were to increase, would it take longer or shorter to make a toy?

H is the number of hours in the day. If workers work more hours, the number of toys will increase.

The question says that workers get training to work more efficiently. Which factor will be affected by this? Training doesn't affect the number of workers, so *w* is out. Training doesn't affect the size of the toy. *V* is out. Nor does training affect the number of hours, so *h* is out.

The only possible answer is choice B, the constant *C*. Such a constant by nature reflects factors that are not specified by the other variables. But you don't need to know that! All the other answer choices are out.

---

8. The expression $15x^2 + 6x + 3$ can be divided by the expression $cx + 3$, where c is a constant and $cx + 3$ is not equal to zero. The result is $5x - 3$ with a remainder of 12. What is the value of *c*?

A) 3

B) 6

C) 15

D) 30

Explanation of 8:

The first thing that comes to your mind might be to factor $15x^2 - 13x + 28$, getting something like (ax+c)(bx+d). A little quick experimenting will show you that the expression can't easily be factored this way.

Orient yourself to the original set up so that you can see the relationships. What the problem tells us that $15x^2 + 6x + 3$ can be divided by $(cx + 3)$ and that it goes in $(5x - 3)$ times with 12 left over. You can write this as

$$\frac{15x^2 + 6x + 3}{cx + 3} = (5x - 3) + \frac{12}{cx + 3}$$

Check to see if the expression above matches the wording of the problem in the way that you understand it. The remainder of 12 is expressed over the $cx + 3$ in the same way that if you divide 9 by 4, you get 2 with a remainder of 1, which represents 1 over 4.

Another way to express the same relationship is that $(cx+3)$ times $(5x-3)$ plus another 12 added on equals the original expression (on the left).

$$15x^2 + 6x + 3 = (cx+3)(5x-3) + 12$$

If you can figure out this relationship from the original problem, you are half way to the solution. This is not an easy relationship to pin down.

Now there are two intuitive ways to get to the answer. Notice that there are two unknowns in the problem: x and c. c is the value we are trying to find. One good strategy is to assign a value to x, so that you are only dealing with numbers, not with variables. The correct value of c will work with any value of x. However, it is usually best to stay away from assigning 0 or 1 because of their special properties. Let's try letting x equal 2.

Starting with the equation above and replacing x with 2, you get:

$$15*2^2 + 6*2 + 3 = (c2+3)(5*2-3) + 12$$
$$15*4 + 6*2 + 3 = (c2+3)(10-3) + 12$$
$$60 + 12 + 3 = (2c+3)7 + 12$$
$$75 = 14c + 21 + 12$$
$$75 = 14c + 33$$
$$75 - 33 = 14c$$
$$14c = 42$$
$$7c = 21$$
$$c = 3$$

Choice A is the correct answer.

The second way to solve this problem involves a bit of a short cut. Without assigning a value to x, you can take the original equation and multiply out the right side.

$$15x^2 + 6x + 3 = (cx+3)(5x-3) + 12$$
$$15x^2 + 6x + 3 = 5cx^2 + 15x - 3cx - 9 + 12$$

The expression on the right side is complex and intimidating. The trick is that the factor on x^2 has to equal 15, as it is on the left side of the equation. For $5c$ to equal 15, c must be 3. You will see problems on your test that require you to use this strategy.

9. The expression $\sqrt[7]{x^{16}}$ is equivalent to which of the following?

A) $x^{\frac{7}{16}}$

B) $x^{\frac{16}{7}}$

C) $x^{\frac{4}{7}}$

D) $x^{\frac{7}{4}}$

Explanation of 9:

This question tests your understanding of various formats of exponents. Here are three examples:

$$10^3$$

$$10^{\frac{1}{3}}$$

$$10^{-3}$$

You are probably clear on what 10^3 means. The other two can be difficult to remember. Here are the two options:

$$\frac{1}{10^3}$$

$$\sqrt[3]{10}$$

Which one is which? It's the opposite of what it might seem. $10^{\frac{1}{3}}$ has a fraction as the exponent but it is **not** equivalent to the expression with the fraction ($\frac{1}{10^3}$).

Simply remember that the fraction exponent goes with the radical (root) expression and the negative exponent goes with the fraction expression.

$$10^{-3} = \frac{1}{10^3}$$

$$10^{\frac{1}{3}} = \sqrt[3]{10}$$

In the problem above you have to go in the other direction. You are given an expression that includes a positive integer exponent and a seventh root. You are asked to rewrite it with just an exponent. x^{16} already contains the exponent. You only need to decide how to write the seventh root as an exponent. Test yourself. Is it $\frac{1}{7}$ or -7? The one that's **not** a fraction (-7) gives an answer with a fraction. We're not looking for an answer with a fraction. So $\frac{1}{7}$ is what we need.

$$x^{\frac{16}{7}}$$

Is that hard to remember? Consider that you can break that up into $x^{\frac{1}{7}}$ times itself 16 times. $\sqrt[7]{x} * \sqrt[7]{x}$ etc.

There is a reason for the way these are written. You don't have to know the reason if you remember that:

    1. There are only two confusing exponents – one with a fraction and one with a negative.
    2. One represents a fraction with the power in the denominator and one represents a root.
    3. The expression with the fraction in the exponent does **not** represent the expression that contains a fraction.

However, if you are interested in the explanation of why it works this way, read the box on the next page.

Why Weird Exponents Work The Way They Do

Negative exponents. Consider the expression below.

$$\frac{10^3}{10^2}$$

You can easily understand this as:

$$\frac{10*10*10}{10*10} = \frac{10}{1} = 10^1$$

The two 10's in the denominator cancel out two of the 10's in the numerator. In other words the three on the top are reduced by the two on the bottom: 3 – 2. This is the same as taking the exponent in the numerator minus the exponent in the denominator.

What happens if the exponent on top is smaller than the exponent on the bottom? You would end up with a negative exponent. Try flipping the example above.

$$\frac{10^2}{10^3} = \frac{10*10}{10*10*10} = \frac{1}{10} = 10^{-1}$$

Does this help explain why a negative exponent means that the number with exponent is in the denominator?

Fractional exponents. This is a little more difficult of a concept. One way to understand it is to consider what it means to "halve" something. If you halve an apple, you cut it into two equal pieces. Piece one plus piece two equal one whole apple, and the two pieces are exactly the same size. So $a + a = b$.

What happens if we try to apply this concept to multiplication instead of addition? What if we take a number, say 25, and break it into two equal pieces such that when you multiply them together, you get 25? So, $a * a = 25$.

This is the definition of a square root. $\sqrt{5} * \sqrt{5} = 25$

Thus, 5^2 asks you to "double" the number five via multiplication: $5 * 5$.

In the same way $5^{\frac{1}{2}}$ asks you to find the number that is "half" of the number 5 in terms of multiplication: $\sqrt{5}$.

10. A factory for refining molasses has a storage tank that is a right circular cylinder. The tank has a diameter of 20 feet and is 30 feet high. A new, larger tank is constructed to hold more molasses. The new tank has a diameter of 60 feet but is only 10 feet high. How much more molasses does the new tank hold compared to the old tank?

A) The new tank holds 9 times more than the old tank.

B) The new tank holds 6 times more than the old tank.

C) The new tank holds 3 times more than the old tank.

D) The new tank holds the same amount as the old tank.

Explanation of 10:

This problem asks you to compare the volumes of two different tanks (right circular cylinders.) A simple way to remember how to calculate such a volume is by thinking of the tank as a stack of pancakes. Each pancake is one foot high.

The base of the tank is a circle but a circle doesn't have volume. It is two dimensional and only has area. If the area of a circle at the base of a tank were, for example, 35 square feet, the first pancake (a "circle" that is one foot high) would contain 35 cubic feet of volume. If the tank were 10 feet high, it would hold ten pancakes, for a total of 350 cubic feet. If the tank were twice as high, it would have twice the number of pancakes and thus twice the volume. If the tank were three times as high, it would have three times the number of pancakes.

In this problem the first tank was 30 feet high and the new tank was one third of that – 10 feet high. The new tank would hold one third the number of pancakes. But each pancake is bigger than the pancakes in the first tank.

The new tank has a diameter that is three times that of the old tank. This is a **trap**! You might think that the new tank is 1/3 the height but 3 times the width, so the volumes would be the same. Here's why this isn't true.

The important factor is the volume of one pancake, which depends on the area of the base. It is true that if the area of the base of the new tank were three times the area of the base of the old tank, that would make up for a height that is one third of the old tank. However, they didn't say that the area of the new tank was three times that of the original. They said that the diameter was three times greater. Why isn't that the same?

How do you calculate the area of the base (a circle)? The formula is πr^2. The radius, r, is half the diameter. For the old tank, $r = 10$. The area of that base is 100π. For the new tank the radius is 30. The area of that base is 900π.

Even though the radius of the new tank is three times that of the old one, the area of the base (and thus the volume of each pancake) is 9 times greater. This is because the radius is squared.

You don't actually have to calculate the volumes of the two tanks. If you call the volume of the original tank O, then the new tank is:

$\frac{1}{3} O$ (because it has one third the height) * 9 (because the radius is three times larger and thus the volume is 3^2 times larger).

$\frac{1}{3} * O * 9 = 3 O$

Choice C is correct.

You finished Day 14! How did it go on these five questions?

Number of questions you got right on your own: _____

Types of problems or patterns you need more work on: _____

_____

How much new did you learn from these questions? ☐ Important tools! ☐ Some tools ☐ Not too much

Day 15, Questions 11-15

11. If $x^2 + 2xy + y^2 = z^2$ and $x^2 - 2xy + y^2 = w^2$, which of the following is equivalent to wz?

A) $x^4 + 2x^2 y^2 + y^4$

B) $x^2 + 2xy + y^2$

C) $x^2 - 2xy + y^2$

D) $x^2 - y^2$

Explanation of 11:

There is a trick to this and it's something you probably should simply memorize. Consider these three expressions:

$$(x + y)(x + y)$$

$$(x - y)(x - y)$$

$$(x + y)(x - y)$$

The first two give you squares: $(x + y)^2$ and $(x - y)^2$. Third expression is not a square but it has some special properties.

You should practice expanding (FOILing) the two squared expressions so that you can recognize them.

$$(x + y)(x + y) = xx + xy + yx + yy = x^2 + 2xy + y^2$$

$$(x - y)(x - y) = xx - xy - xy + yy = x^2 - 2xy + y^2$$

Notice that in the first expression, the two *xy* terms are positive. In the second, they are both negative. In the third expression, one is positive and one is negative, so they cancel each other out and all that is left is the two squared terms and one of those is negative.

$$(x + y)(x - y) = xx + xy - xy - yy = x^2 - y^2$$

These are common patterns that will appear on your test, so memorize them! In this problem the first expression - $x^2 + 2xy + y^2 = z^2$ contains the perfect square $(x + y)^2$ so $z = (x + y)$. Likewise the second expression tells us that $w = (x - y)$. The problem then asks us to multiply *wz*, which is the same as $(x + y)(x - y)$ and thus equals $x^2 - y^2$. Choice D is correct.

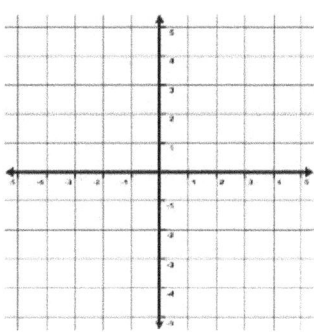

12. On an xy-plane the point $(0,3)$ could be on which of the following figures?

A) $y = x + 3$

B) $y = 0$

C) $y = \dfrac{3}{x}$

D) $y = \dfrac{x^2}{3}$

Explanation of 12:

This problem is easier than it may seem. You don't need to graph the equations. The key here is to recognize that the point $(0,3)$ means that when x is 0, y must be 3. Plug 0 in for x in the answer choices.

For choice A, when x is 0, y is 3. This works and must be the correct answer. Double check by doing the others. For choice B, there is no x and y must always be 0, so it can never by 3. For choice C, if x is 0, the expression is undefined. For choice D if x is 0, y is 0.

$$3(y + 5) = 6x + 15$$

13. If (x, x^2) is a point on the line described above, what is the product xy at that point?

A) 18

B) 9

C) 8

D) 2

Explanation of 13:

You might feel intimidated by the algebra and by the x^2 and by the fact that they are giving you a point on an xy graph. Your first step is to ignore the intimidation and just start with something you are confident in.

Let's first rewrite the equation. It would be helpful to end up with an equation in the form of $y = ax + b$. With that equation you can assign values to x and then find out what the y would be.

Expand the left side of the equation:

$$3y + (3 * 5) = 6x + 15$$

or

$$3y + 15 = 6x + 15$$

You can use your intuition to see that if Thing 1 plus 15 is the same as Thing 2 plus 15, then Thing 1 must be the same as Thing 2. Pictorially, it looks like this:

$$\heartsuit + 15 = \clubsuit + 15$$

$$\heartsuit = \clubsuit$$

So:

$$3y = 6x$$

Now simplify this:

$$y = 2x$$

Great! You've simplified the equation. At this point you might have forgotten what the problem is actually asking (this happens a lot!) so be safe and review the question stem.

If (x, x^2) is a point on the line described above, what is the product xy at that point?

There are a couple ways that you can deal with this without getting lost in too much algebra. One way is to note that x^2 is really the y coordinate of the point (x, x^2). If $x^2 = y$, then we can replace the y in the equation with x^2.

$$x^2 = 2x$$

Can you look at that equation intuitively?

$$x * x = 2 * x$$

Is it clear that x is 2?

$$2 * 2 = 2 * 2$$

You're almost there but not quite. Check the question stem again. They are not asking what x is. They are asking for the product:

$$xy$$

Plug in the values you've figured out. Remember that x is 2 and y is 2^2.

$$2 * 2^2 = 8$$

The answer is choice C.

If the above doesn't work too well for you or if you are interested in seeing an alternate tool, consider this. Once you have $y = 2x$, you can make a table to find various values for y, given certain values for x. This helps you keep track of all the variables you need to know.

   Variable 1. x. Column 1. You will assign values to this.
   Variable 2. y. Column 2. Calculate this by using $y = 2x$.
   Variable 3. x^2. Column 3. You are looking for the one value of x for which y (second column) equals x^2 (third column.)
   Variable 4. xy. Column 4. For the value of x where column 2 equals column 3, the answer you are looking for is xy, column 4.

This kind of table is an amazing way to organize all of the information in the problem without having to use any algebra!

It's a good idea to start with something simple. Let's make x equal to 1 and then increase it as we go along.

x	y (=2x)	x^2	xy
1	2	1	2
2	4	4	8
3	6	9	18
4	8	16	32

As soon as you get to $x = 2$, you can see that y is the same as x^2. This is what we're looking for and so the number in the last column, xy, must be the answer. In the example above, we tried 3 and 4 as a double check. As x got larger than 2, y got further and further away from x^2. This shows us an important trend. There's no point trying larger numbers for x. That would just make y get even further from x^2.

For this problem the two methods are equally effective. The second method is a great example of the power of using a table to organize relationships between variables.

$$(3 + 5i) + (4 + 7i)$$

14. Where $i = \sqrt{-1}$, the above sum is equal to

A) $19i$

B) $19i^2$

C) $12 + 35i$

D) $7 + 12i$

Explanation of 14:

Remember that i is used to express imaginary numbers. This is because $\sqrt{-1}$ has no real value. You can't multiply any real number by itself to get -1 because the product of two negatives is a positive number and the product of two positive numbers is a positive.

Don't let the fact that i is imaginary bother you. Even though it **is** imaginary, if you have one i and then get another one, you have $2i$.

Rearrange the original equation so that you have similar factors near each other.

$$3 + 4 + 5i + 7i$$

$$7 + 12i$$

Voila!

15. Which graph below correctly represents the equation $y = 3x - 2$?

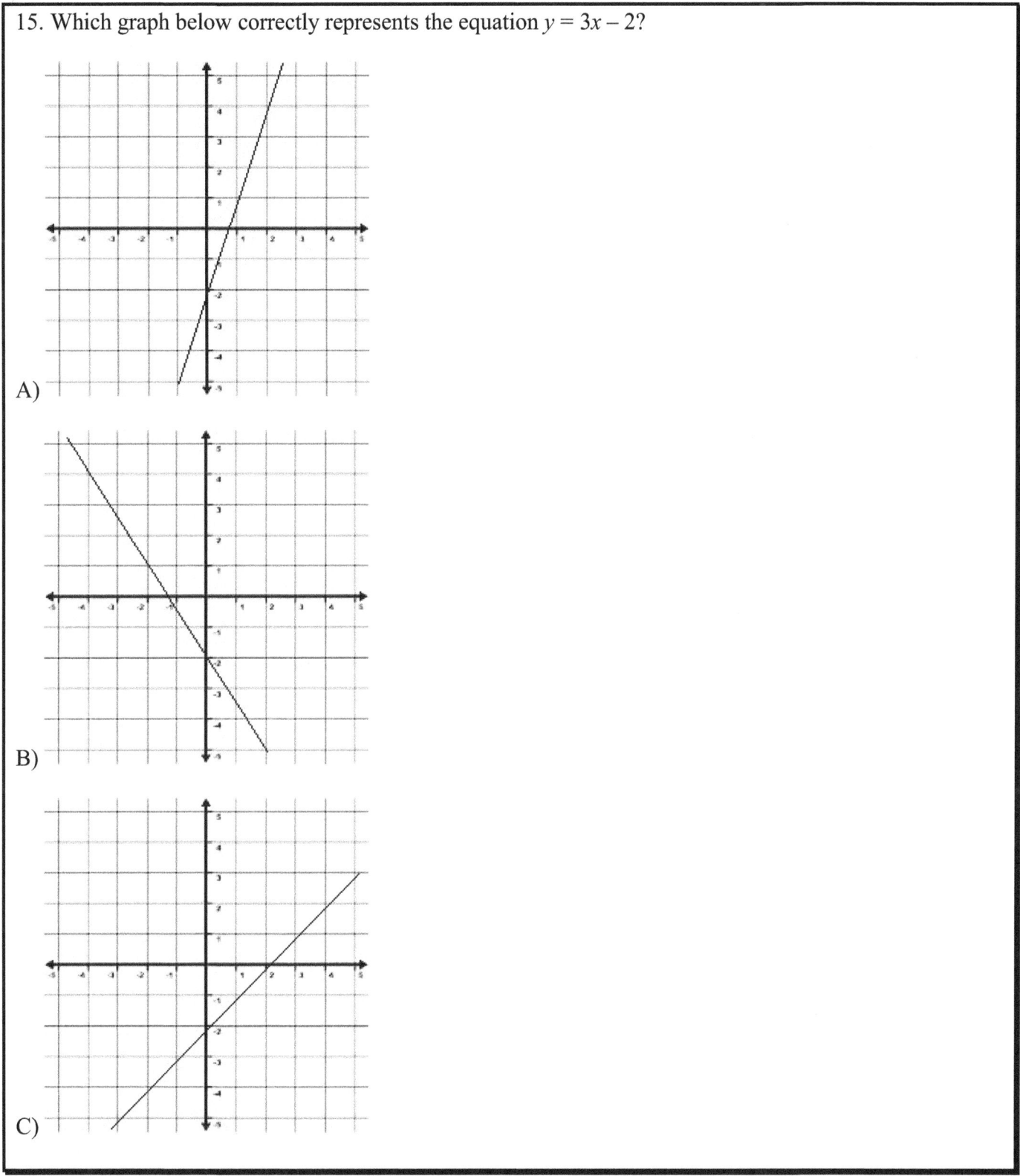

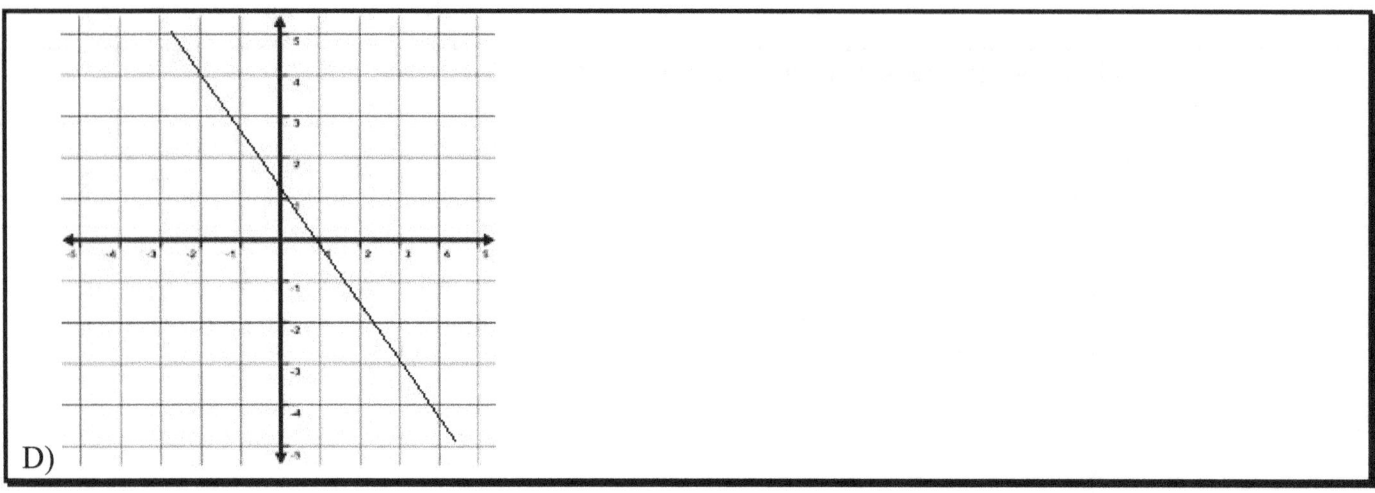
D)

Explanation of 15:

The easiest way to answer this question is to find one or two points that you know for sure are on the line. And the easiest way to do that is to set x to zero and solve for y and see which graphs are eliminated. When x is 0:

$$y = 0 - 2 = -2$$

Where on the graph x is equal to zero? Of course x is zero at the origin but what is the set of all points for which $x = 0$? Do you see that it would be the y-axis? Good!

So, when x is 0, y is -2, which is the point $(0,-2)$. We can eliminate any graph with a line that does not cross the y axis at -2. In choices A, B, and C, the lines do go through $(0,-2)$. However, in choice D the line does not. Choice D is out.

The next easiest point to find is when $y = 0$.

$$0 = 3x - 2$$
$$0 + 2 = 3x - 2 + 2$$
$$2 = 3x$$
$$x = \frac{2}{3}$$

This gives us the point $(\frac{2}{3}, 0)$. When y is zero, the line is passing through the x axis. We can now eliminate any answer choices in which the line does not pass through the x axis at $\frac{2}{3}$. Choice A looks close. In choice B the line passes through the x axis at a negative point. Choice B is out. In choice C the line passes through $(2,0)$. Choice C is out. Choice D was already eliminated, so we've confirmed that choice A is the correct answer.

You finished Day 15! How did it go on these five questions?

Number of questions you got right on your own: _____

Types of problems or patterns you need more work on: _____

_____

How much new did you learn from these questions? ☐ Important tools! ☐ Some tools ☐ Not too much

Day 16, Questions 16-20

Instructions: For questions 16-20 there are no multiple choice answers. You must solve the problem on your own. No answers are negative. Some problems have more than one correct answer. You only need to put down one correct answer. Your answer can be a fraction or decimal.

$$3x + 5y = 1$$
$$2x - 3y = 7$$

16. Given the above system of equations, what is the value of x?

Explanation of 16:

The trick to solving this problem is to add or subtract the two equations so that one variable drops out. To do that, you have to rewrite one or both equations in a different but equal form, usually by multiplying both sides of the equation by the same number (as long as it isn't zero!)

Because we are solving for x, let's try to make the y's drop out. Theoretically, you could multiply $3y$ times a fraction to get $5y$ but that's cumbersome. So in both equations let's try to turn the y factors into $15y$.

$$3(3x + 5y) = 1*3$$
$$5(2x - 3y) = 7*5$$

$$`9x + 15y = 3$$
$$10x - 15y = 35$$

Now we can add the equations.

$$19x = 38$$
$$x = 2$$

Chapter 5. SAT®-style Non-Calculator Questions with Explanations

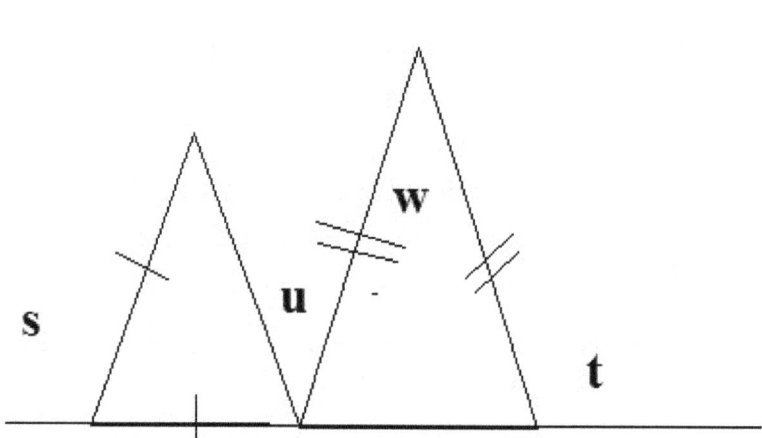

Note: Figure not drawn to scale.

17. The two triangles shown above are isosceles. Angle *s* is 104°. Angle *u* is 48°. What is the value of *w* + *t*?

Explanation of 17:

To organize your work, let's label the unlabeled angles in the two triangles. In the left triangle, let's call the bottom left angle *a*, and the top angle *b*, and the angle in the bottom right corner *c* (*b* and *c* are equal because they are opposite the two equal sides of the isosceles triangle.) In the right triangle, let's call the bottom left angle *x* and the angle on the bottom right *y*.

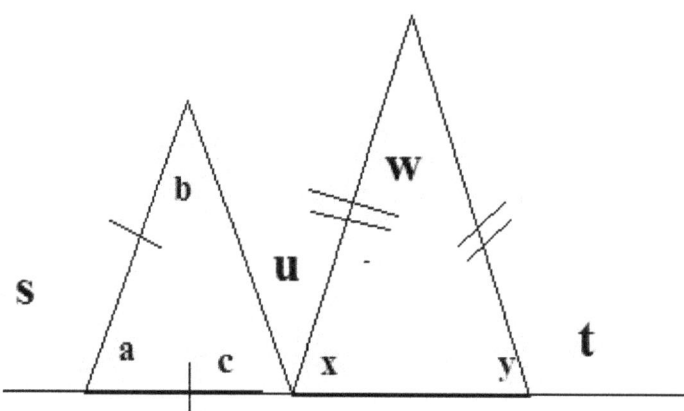

If angle *s* is 104, then *a* must be 76°, because the two angles together equal a straight line, 180°. Because angle *a* is 76°, the remaining angles must total 104, as all three have to add up to 180. The other two angles are equal, so they must be 52° each.

168 Chapter 5. SAT®-style Non-Calculator Questions with Explanations

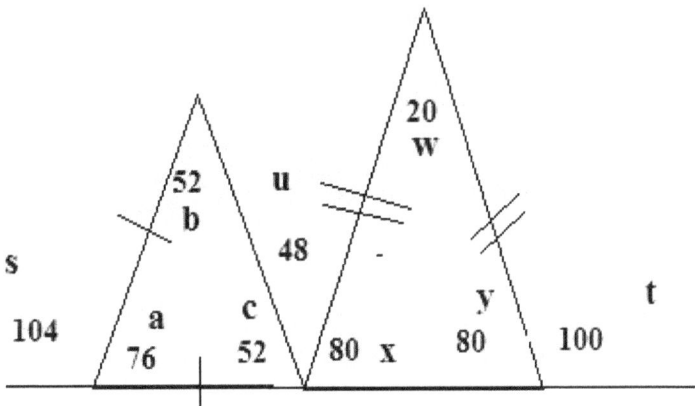

Because angle *c* is 52°, we can add it to *u*, which we are told is 48°. The total is 100°. Angle *x*, in the triangle on the right, must then be 80° to make up the 180° of the straight line. Angles *x* and *y* are equal and thus add up to 160°. Angle *w* must then be 20°.

Angle *t* and angle *y* must total 180°. We figured out that *y* is 80°, so *t* must be 100°. The question asks us to add *t* and *w*.

$$100 + 20 = 120°$$

The most important strategy in this question is to **label** everything! Then you can figure out step by step what each angle is.

$$\frac{9+5\sqrt{x}}{x} - \frac{5}{\sqrt{x}} = \frac{c^2}{x}$$

18. In the above expression, *c* is a constant and $x \neq 0$. What is the positive value of *c*?

Explanation of 18:

Your first step is to notice how intimidating this problem looks. Your second step is to take a breath and consider what you can do. There are two things in the equation that make it complicated: x in the denominator and $\sqrt{x}$. To simplify x in the denominator, you can multiply by x. To simplify $\sqrt{x}$, you can multiply by $\sqrt{x}$ or divided $\sqrt{x}$ by $\sqrt{x}$.

Your third step is to consider which of these two paths you want to take. It's often hard to know without actually trying them. However, we can see that multiplying both sides by $\sqrt{x}$ would still leave some factors with $\sqrt{x}$ and so might not really simplify things. Let's try the other path.

We'll multiply both sides of the equation by x. To do this, it's important to know that x cannot equal zero because multiplying both sides by zero just gives us zero everywhere! Do the algebra carefully.

$$x\left(\frac{9+5\sqrt{x}}{x} - \frac{5}{\sqrt{x}}\right) = x\left(\frac{c^2}{x}\right)$$

$$x\left(\frac{9+5\sqrt{x}}{x}\right) - x\left(\frac{5}{\sqrt{x}}\right) = c^2$$

In the line above, notice that on the right side, the x on top cancels the x on the bottom. On the left side, in the first expression the x's will also cancel.

$$(9+5\sqrt{x}) - \frac{5x}{\sqrt{x}} = c^2$$

It is a little frustrating at this point to still have $\sqrt{x}$, but problems are often simpler than they look, so let's just keep pushing forward. We can simplify the second expression on the left. The trick is that x can be broken down into $\sqrt{x} * \sqrt{x}$.

$$(9+5\sqrt{x}) - \frac{5\sqrt{x} * \sqrt{x}}{\sqrt{x}} = c^2$$

Aha! The $\sqrt{x}$ in the denominator is cancelled by one of the $\sqrt{x}$'s in the numerator. Without knowing we were doing it, we have eliminated a denominator simply by pursuing ways to simplify the equation.

$$9 + 5\sqrt{x} - 5\sqrt{x} = c^2$$

The two terms with $\sqrt{x}$ cancel each other out. So as it turns out, we didn't need to worry about getting rid of them. The problem was set up to make it possible to simplify. You can count on the test to often make it easier for you than it first appears! So keep trying something.

$9 = c^2$

and therefore, the positive value of c is 3.

19. If $\sqrt[4]{x^y} = 9$, give a possible value for y.

Explanation of 19:

This problem most likely looks complex to most people. Let's break it down. Problems are almost never as difficult as they first look!

The problem involves fourth roots, a variable taken to a variable power, and the combination of the two. It's a good idea to orient yourself to these issues one at a time to keep things simple. Let's look at a simple example of a fourth root in order to understand it more clearly. In fact let's start by orienting to square roots.

$$\sqrt{16} = 4$$

Consider what this means.

$$16 = 4 * 4 \text{ or } (\sqrt{4*4} = 4)$$

Now consider how this works with a variable under the radial.

$$\sqrt{x} = 5$$

What is *x*?

$$\sqrt{5*5} = 5$$

or

$$25 = 5 * 5$$

In words we can say, "If the square root of a number is 5, then that number is 5 times 5." You can also simply square both sides of $\sqrt{x} = 5$ to get $x = 25$.

If that's clear, then consider a fourth root.

$$\sqrt[4]{32} = 2$$

Is it clear that this means that 2 * 2 * 2 *2 equals 32?

Chapter 5. SAT®-style Non-Calculator Questions with Explanations   171

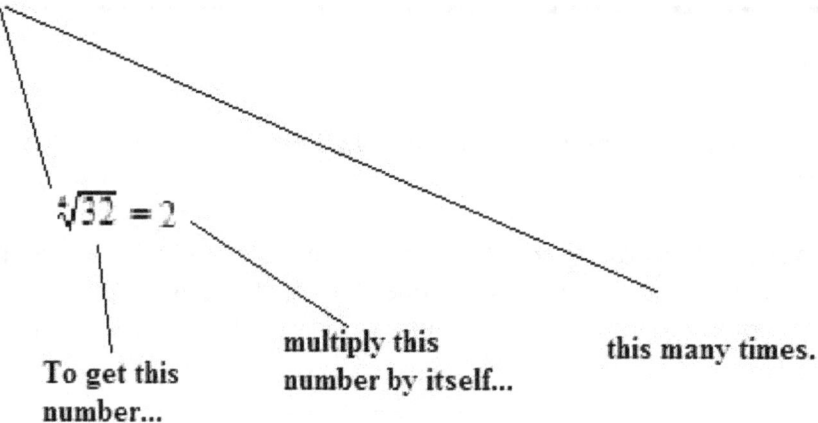

Go back to the original equation.

$$\sqrt[4]{x^y} = 9$$

This means that $9 * 9 * 9 * 9 = x^y$. If that's not completely clear, take each side to the fourth power.

$$x^y = 9^4$$

The problem asks you to give a possible value for *y*. This means that there is more than one correct answer. The above equation tells you that if *x* is 9, *y* would be 4, so 4 is a correct anwer. That's all you need.

In case you're interested in what other answers might be, let's break the equation into its prime factors:

$$3*3*3*3*3*3*3*3 = x^y$$

You can play around with this a bit. The most obvious new possibility is 3^8. But you can also regroup the 3's in various ways.

$$(3*3*3*3)*(3*3*3*3) = 81^2$$

You can also take the original 9 * 9 * 9 *9 and multiply it out. It comes to 6561, so another solution is simply 6561^1

Remember that the question is asking for *y*, the exponent. So correct answers would be 1, 2, 4, or 8.

The lesson to be learned from this question is to orient yourself to the concepts in a question by using simple examples. And don't get intimidated by a complex set of concepts. Deal with them one at a time.

20. In a given right triangle, one angle has a measure of $t°$, such that $\sin t° = \dfrac{2}{3}$. What is the value of $\cos(90° - t°)$?

Explanation of 20:

This is a trick question! Don't let the sines and cosines worry you. Our best tool is to break things down one at a time. That keeps it simple and doable. First, let's draw the triangle.

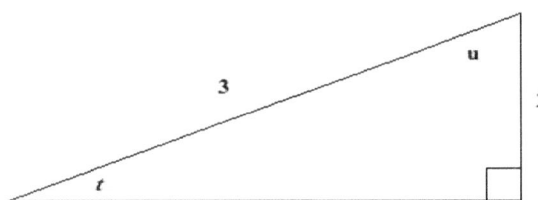

We like to use this orientation as a standard form, with the right angle in the lower left corner, to orient to the trigonometric functions. Starting from angle t, the sine is the opposite over the hypotenuse: in this case 2 over 3.

So far so good. The problem asks about the cosine of a different angle. Where is the angle $(90° - t°)$? It's not t and it's not the right angle. It has to be the angle at the top right of the triangle, which we have decided to call u. You can also figure that out by considering that all three angles total 180. If you take out the 90° angle, the other two must equal the remaining 90. So u is $90° - t°$. In other words if t and u together equal 90, then u is whatever number of degrees it takes to get from the value of t up to 90.

Is it clear now that the question is asking for the cosine of u? In order to more easily see the trigonometric relationships, let's redraw the triangle in a standard form.

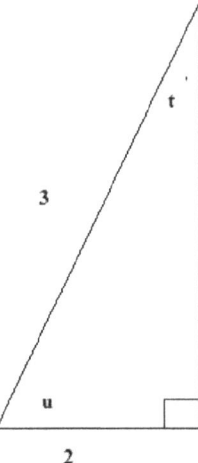

Chapter 5. SAT®-style Non-Calculator Questions with Explanations

Whereas sine is the opposite side over the hypotenuse, the cosine is the adjacent side over the hypotenuse. That gives us 2 over 3 – exactly the same two sides that we used to get the sine of *t*. It turns out that the sine of one angle in a right triangle is the same as the cosine of the complementary angle.

Let's do a quick visual review of the main trigonometric functions.

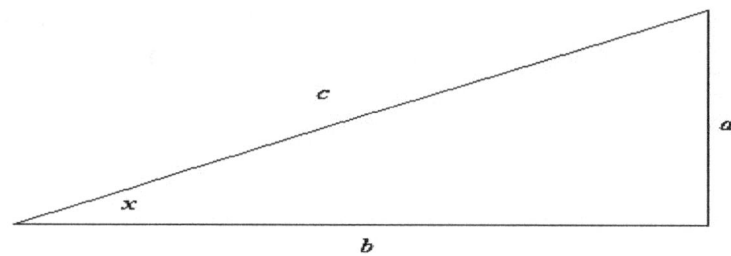

The tangent of *x* is the side opposite *x* over the side adjacent to $x = \dfrac{a}{b}$.

The cotangent is the inverse of the tangent $= \dfrac{b}{a}$

Tangent and cotangent are the only trigonometric functions that do **not** include the hypotenuse. They are also the only two functions for which the "co" function is simply the reciprocal (upside down!) of the other. In other words sine and cosine are **not** the reciprocals of each other.

Sine: In the above diagram, the sine of *x* is the opposite over the hypotenuse $= \dfrac{a}{c}$.

Cosine: The cosine is the adjacent side over the hypotenuse $= \dfrac{b}{c}$.

The remaining functions are secant and cosecant. These are the inverses of the cosine and sine but not in the way you might think. The "co's" don't go together!

Secant is the inverse of cosine. Cosecant is the inverse of sine. Do you have a good way of remembering these not-so-intuitive relationships?

You finished Day 16! How did it go on these five questions?

Number of questions you got right on your own: _____

Types of problems or patterns you need more work on: _____

_____

How much new did you learn from these questions? ☐Important tools! ☐Some tools ☐Not too much

Congratulations on finishing Chapter 5!

Chapter 6. SAT®-style Calculator Questions with Explanations

The questions in this chapter are written and formatted in the style of questions on the SAT® exam. They represent the most common patterns of math questions found on both the ACT® and SAT® exams. You are allowed to use a calculator on this type of SAT® section. However, using a calculator is not always the most accurate way to get an answer. Another SAT® math section does not allow the use of a calculator.

Try to work each problem on your own before reading the explanation. Take as much time as you need. It can be helpful to give yourself 15 or even 30 minutes or more to work on a problem. The longer you work on it, the more you can learn. The explanations are designed to be simple and intuitive. Nevertheless, you will probably find some explanations challenging. Stick with it. Experiment with it. Work on it with a friend.

In some problems you might find that the intuitive strategies seem unnecessary and that the problem can be easily solved with standard math tools. We suggest that you still study the intuitive tools for that problem. You may need those tools on another problem.

Your primary focus is to learn new ways of thinking about mathematical relationships. It does not matter that much whether you get a question right or wrong now. The patterns in this chapter are the patterns you will see on your test. Study them, learn them, and get comfortable with intuitive tools for solving them.

You can follow the daily assignments – about five questions per day – or you can do more or fewer questions per day. At the end of each day's assignment you can evaluate how you did.

Day 17, Questions 1-5

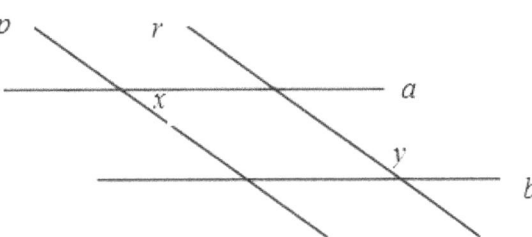

1. In the above diagram lines p and r are parallel. Lines a and b are parallel. If the measure of $\angle y$ is 105, what is the measure of $\angle x$?

A) 60°

B) 75°

C) 90°

D) 105°

Explanation of 1:

When parallel lines intersect, they create equivalent angles, which means that in the above diagram, all the "big" angles are the same and all the "small" angles are the same. One way to understand and remember this is to imagine that you are taking line *p* and moving back and forth along lines *a* and *b*. As long as *a* and *b* remain parallel to each other and as long as you keep line *p* at the same angle, the angles that *p* forms with the other lines stay the same.

To solve this problem and find *x*, you simply need to find out what the "small" angles are. Because *y*, the "big" angle, is 105, the "small" angle to the left of *y* must be 75 in order to make the full 180° of a straight line. If you need to use your calculator to figure that, that's fine. Remember that you can make errors in entering information on your calculator, so also practice doing the calculation on paper. You may find that you are more accurate that way.

Not too accurate at subtractions? There are intuitive tools that you can use. Consider trying to figure the difference between 180 and 105. Here's one way to do it on paper without having to subtract.

Start	Add	New total
105	5	110
110	70	180
Total added:	**75**	

We started with 105 and just added numbers a little at a time in a way that you can do the math in your head. Adding 5 to 105 gives 110. Adding 70 to that gives 180. We added 5 and then 70 so we added 75. Simple.

If you are not able to make a leap directly from 110 to 180, just keeping adding 10 until you get to 180. Then count up all the numbers you added. Adding is less prone to error than subtracting and adding simple numbers is easier than adding more complex numbers.

All the small angles, then, are 75, so that is also the value of *x*. Choice B is correct.

$$\frac{(x+3)(x-3)}{|x|+3}$$

2. For what value(s) of *x* is the above fraction undefined?

A) 3 only

B) –3 only

C) 3 and -3

D) The fraction is defined for all values of x

Chapter 6. SAT®-style Calculator Questions with Explanations

Explanation of 2:

Consider what your first step should be. Don't be tempted to start multiplying everything out unless you have a good reason to do so.

In this question a good first step is to orient yourself to the question stem. What does it mean for a fraction to be undefined? If you remember that an undefined fraction is one in which the denominator is zero, you're in luck. You won't need to do anything with the numerator. You simply have to find out what value of x would make the denominator zero.

$$|x| + 3 = 0$$

You can now use your intuition. What value would the ♦ symbol below have to have to make the equation work?

$$♦ + 3 = 0$$

Is it clear that ♦ would have to be –3? Now we have:

$$|x| = -3$$

What is x? Consider the definition of $|x|$, read "absolute value of x." Here's a good way to remember absolute value of a number. On a number line the absolute value of a point is the distance that the point is from zero. And in this situation distance can **only** be a positive number. Negative 3 and positive 3 are both 3 units from zero.

If an absolute value can only be a positive number, how can there be an x that has an absolute value of –3? There can't!

The correct answer is choice D. The fraction is defined for **all** values of x.

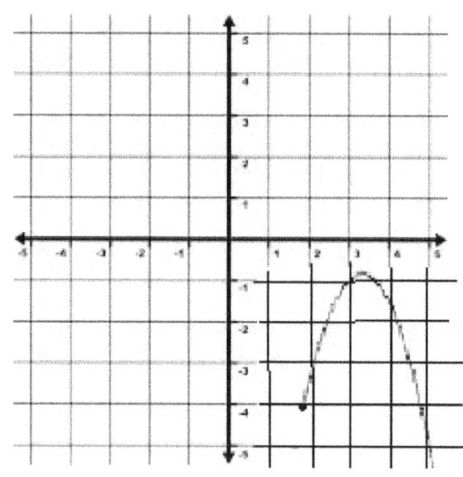

3. In the figure above plotted on the xy-plane, which of the following relationships between x and y must be true?

A) $y^2 > x$

B) $y > x$

C) $x > y$

D) $x^2 < y$

Explanation of 3:

Approach this by orienting yourself to the most basic information. What do you know about the x values and the y values in the graph?

The x values are all positive, ranging from about 1.5 to above 5. The y values are all negative, ranging from a little less than –1 to beyond –4 and –5.

This may be all you need to know to answer the question, so check the answer choices. Avoid making the problem more complex than it is!

In choice A, a negative number squared could be larger than x but does not have to be. This is a *must* question. The correct answer **must** always be true. Choice A could be true but does not have to be.

In choice B, y must always be less than x, because y is negative and x is positive. Choice C is the reverse of choice B. Yes, x must be larger than y. This has to be the answer. Just to double check yourself, make sure you can eliminate choice D. x squared is positive and cannot be less than y.

4. A rare tapestry measuring 2.5 meters by 20 meters was examined by ten experts. Each expert examined a section of the tapestry measuring 1 meter by 0.5 meters. The experts counted the number of threads in their section. The table below displays the results.

Expert	Threads	Expert	Threads
1	1240	6	1298
2	1385	7	1345
3	1198	8	1301
4	1322	9	1284
5	1269	10	1326

If the number of threads is relatively consistent throughout the tapestry, which of the following is the most likely estimate of the total number of threads in the tapestry?

A) 1,300,000

B) 130,000

C) 13,000

D) 1,300

Explanation of 4:

The key to this question is to take the time to orient yourself carefully to the information. What does the question require you to do? To be clear on this, let's reread the part of the setup that poses the question. If the sentence seems so wordy that it's hard to follow, try simplifying the language for yourself. We might say:

"If the number of threads is pretty consistent, about how many total threads are there in the whole thing."

Now it's clear that we need the total number of threads. Take stock of what information we have. We have ten different counts of the number of threads in part of the tapestry. The ten experts are pretty close to each other in their count. How do we get the total number of threads?

Stepping back from the problem a bit, we can make a simple example to understand the problem. If we knew how many threads were in half the tapestry, we could double it to get the total. If we knew how many were in a tenth, we'd multiply it by 10.

There are several steps involved in getting to the answer so we should create a road map of our plan of attack.

    1. Determine what fraction of the tapestry we have information on
    2. Determine an average number of threads for that portion of the tapestry.

Chapter 6. SAT®-style Calculator Questions with Explanations

3. Multiply the number from step 2 by some number, based on step 1, to get threads for the whole tapestry.

So what fraction of the whole tapestry do we know about? A fraction is a ratio of a part to the whole.

$$\frac{part}{whole}$$

Fill in what we know. The part is 1 meter by 0.5 meters. We need to decide what units we want to use to represent the part and the whole. Because we're talking about area, we can use square meters.

$$Part = 1 * 0.5 \text{ square meters} = 0.5 \text{ m}^2$$

$$Whole = 2.5 \text{ m} * 20 \text{ m} = 50 \text{ m}^2$$

$$\frac{part}{whole} = \frac{0.5}{50} \text{ m}^2 = \frac{2}{2} * \frac{0.5}{50} = \frac{1}{100}$$

We've completed step 1 of our plan. The ratio is 0.5 to 50, which is the same as 1 to 100. The sample is one hundredth of the whole tapestry. To get the number of threads for the tapestry, we will multiply the threads in the sample times 100 (step 3.)

Step 2 is to estimate the number of threads in the sample. Hint: look at the answer choices!! They are all based on 1300.

Step 3 is to multiply 1300 by 100. Be careful about how many zeroes you need. The safest way to multiply by hundred is to think of it as adding two zeroes. Add in a comma to separate the thousands and you have:

$$130{,}000$$

Choice B is the answer.

Notice that so far you have not really needed the calculator for these calculator-section questions.

5. Shannon decided to run for two hours on a country road in order to build her endurance. She planned to push herself to continually accelerate as long as she was running on a flat stretch of road. However, if she got to an uphill stretch, she would run at a constant rate and if she got to a downhill stretch, she would decelerate as she ran. If her heart rate went over 120, she would take a short break. Below is the graph of her run.

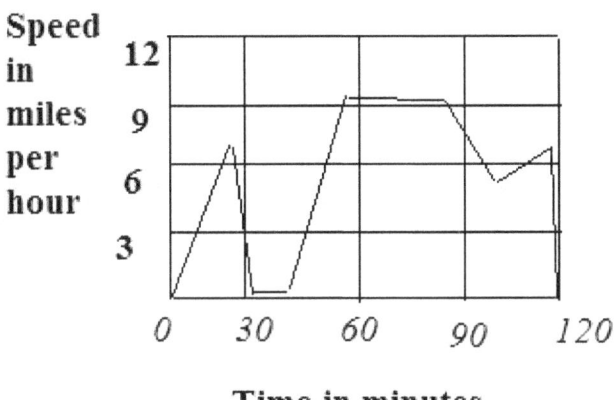

Approximately how long did Shannon run along an uphill stretch?

A) 45 minutes

B) 30 minutes

C) 20 minutes

D) 15 minutes

Explanation of 5:

The key to this question is in understanding how the graph represents the various stages of Shannon's run. First, orient yourself to the units on the graph. The *x*-axis is time in minutes. The *y*-axis is speed in miles per hour.

Next, consider what the stages of her run are and what they would like like on the graph.

        Flat – accelerate; line goes up

        Uphill – steady speed; line stays flat

        Downhill – decelerate; line goes down

High heart rate – stop; line stays flat at zero

Next orient yourself to how each of these stages appear on the graph. Find the parts of the graph in which Shannon is accelerating. The main such areas are from 0 minutes to about 27 minutes and then again from about 40 minutes to about 58 minutes. As an exercise, identify the parts of the graph that correspond to each stage.

The question only requires you to identify the amount of time during which Shannon is running uphill. Uphill is represented by a flat line – meaning that her speed is not increasing. There are two places on the graph in which the line is relatively flat. However, there is a trap here! The first flat area, at about 30 minutes, has a speed of zero. She is **not** running! This is what stage 4 looks like – high heart rate and taking a break. Only the second flat area represents running uphill. It lasts for about 30 minutes and answer choice B is correct.

You finished Day 17! How did it go on these five questions?

Number of questions you got right on your own: _____

Types of problems or patterns you need more work on: _____

_____

How much new did you learn from these questions? ☐ Important tools! ☐ Some tools ☐ Not too much

Day 18, Questions 6-10

Questions 6 and 7 refer to the following information.

Object	Mass (kg)
A	2.3
B	7.8
C	3.6
D	4.7
E	10.5
F	7.2

In physics, force is defined by the equation

$$F = ma,$$

where F is force, measured in Newtons (N), m is mass, measured in kilograms (kg) and a is acceleration, measured in meters per second squared, $\frac{meters}{second^2}$. The chart above shows the mass for six distinct objects.

---

6. For the same rate of acceleration, if the force acting on Object A is 15 N, approximately what would be the force acting on Object D?

A) 4.7 N

B) 7.5 N

C) 7.5 $\frac{m}{sec^2}$

D) 30 N

Explanation of 6:

Your first task is to orient yourself carefully to all of the information in the presentation, especially because you'll need it for two questions. Notice what you have to distinguish and what you don't have to distinguish. You have to distinguish the six objects from each other. You have to distinguish objects from mass.

Also make sure you have oriented yourself to the physics equation. Don't worry too much about whether you understand physics. You don't have to. They are testing math relationships. Consider the relationships in the equation. If m gets bigger, F will get bigger. If a gets bigger, F will get bigger. This may be all you need to know.

The question tells you:

$$2.3 * a = 15 \text{ N}$$

and then asks you to find:

$$4.7 * a = ?$$

The above numbers come from objects A and D, respectively. What is the relationship between 2.3 and 4.7? Do you notice that 4.7 is very close to double?

You might have been tempted to try to solve the first equation for a but it's not necessary. The second equation is simply two times the first one.

$$2 * 2.3 * a = 2 * 15 \text{ N}$$

$$4.6\, a = \sim 30 \text{ N}$$

Choice D is the correct answer.

The equation $F = ma$ is a linear one. If you double m, then F doubles. The same is true for a.

Not all equations have a linear relationship. Consider the formula for the area of a circle.

$$A = \pi r^2$$

If you triple r, the area doesn't triple. It is multiplied by 3^2 (=9) because r is squared. Some problems on the test will try to trick you with this. If the radius of a circle doubles, what happens to the area? It doesn't double. It is increased by 4 (2^2).

---

7. If the force acting on Object E is 10.5 N when the acceleration is $\dfrac{1 meter}{(1 \sec)^2}$, approximately what would be the force acting on Object E if the acceleration was changed to $\dfrac{1 meter}{(3 \sec)^2}$?

A) 1.1 N

B) 3.5 N

C) 30.5 N

D) 94 N

Explanation of 7:

This question does exactly what you were warned about in question 6! See if you can avoid the trap. You remember that when a fraction is multiplied times another number and the denominator is increased, the product gets smaller. Because the new denominator is increased from 1 second to 3 seconds, you may be tempted to multiply 10.5 by one third. However, the seconds are squared, so in fact you have to multiply 10.5 times one ninth, which is between 1.1 and 1.2, so choice A is correct. You don't need the calculator for this. Nine goes into ten once with a little left over. Choices B, C, and D are much too large. The lesson? Don't get hung up on exact calculations when an approximation will do.

$$-4y^2 + 5y + 3$$
$$7y^2 - 8y - 5$$

8. Summing the two polynomials above gives which of the following expressions?

A) $3y^4 - 3y^2 - 2$

B) $28y^4 - 40y^2 - 15$

C) $3y^2 - 3y - 2$

D) $3y^2 - 13y - 2$

Explanation of 8:

This is relatively easy because the problem gives you lots of clues. Look at the answer choices. Two of them include y^4. You are asked to add the two equations. The only way to get a fourth power would be to multiply the two. Choices A and B are out.

Compare choices C and D. Both have $3y^2$, so you don't need to think about whether that is correct. It has to be. Both have –2. Again, you don't need to think about this. The only difference between C and D is –3y as opposed to –13y. Both are negative.

Examine the y factors in the equations. There is +5 and –8. They add up to a negative but both choice C and D have a negative y factor. You have to actually add 5 and -8. If you aren't careful, you might mistakenly add 8 and 5 and believe that the answer has a 13 in it. Double-check your addition by using an intuitive approach. 8 is only three more than 5. If you subtract 8 from 5, you will only go 3 into the negative.

Another intuitive approach is to make a concrete example with something like dollars or apples. I have five apples. I owe you eight. You come to collect and only get five. How many more do I owe you?

Even the sharpest mathematician can make errors in simple calculations when their brain is tired. Double-checking yourself with intuitive strategies can save you from a lot of wrong answers. And using the calculator is not necessarily an intuitive strategy.

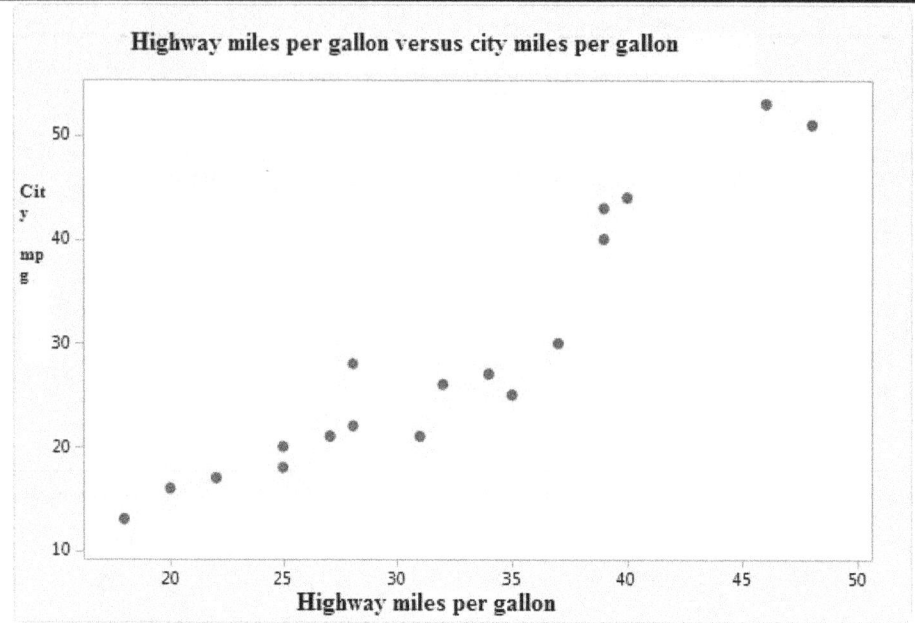

9. The above scatterplot shows highway miles per gallon for various vehicles compared to city miles per gallon. The average of this relationship can be expressed by the equation $c = kh$, where c is city mileage, h is highway mileage, and k is a constant. Which of the following is closest to the value of k based on the above chart?

A) 1.7

B) 1.2

C) 1

D) 0.7

Explanation of 9:

This problem contains some interesting relationships. With any graph, your first task is to orient yourself to it. What do each of the axes represent and what is their relationship. In this graph the *x*-axis represents miles per gallon on the highway. The *y-axis* also represents mile per gallon but in the city. Take a moment to let this make sense to you. It helps if you are aware that cars typically get better mileage on the highway than in the city. (This is because stop and go driving in the city is less fuel-efficient.)

Even if you weren't aware of that fact, you can get an idea of the relationship by examining the graph. It might not be clear at first what each dot represents. A good strategy is to consider just one dot and focus on it. Let's look at the second dot from the left. Examine the data for this dot. The highway mileage is about 20 mpg. The city mileage is about 17 mpg. Does it make sense to you that the dot represents one particular car?

Consider the point on the *x-axis* labeled 25. There are two dots above it. These must represent two different cars. One gets about 18 mpg in the city and the other about 20 mpg.

The above steps are to help you get oriented to the graph. If you felt comfortable with the graph from the beginning, you don't necessarily need to these steps.

Once you are oriented to the graph, the next step is to orient to the question stem. The stem tells you the relationship between the city mileage and the highway mileage. On the average the city mileage is some constant k times the highway mileage. Make sure that makes sense to you before moving on.

By analyzing a few points, you can see that city mileage is consistently less than highway mileage. Therefore, the constant must be less than 1. In other words, if you start with the highway mileage, you have to make it smaller to get the city mileage. Only answer choice D is less than 1.

You're done! The lesson? Don't overwork the problem. Stop periodically to see if you already have enough information to find the answer or get closer to it. Use concrete information. Don't get lost in the abstract.

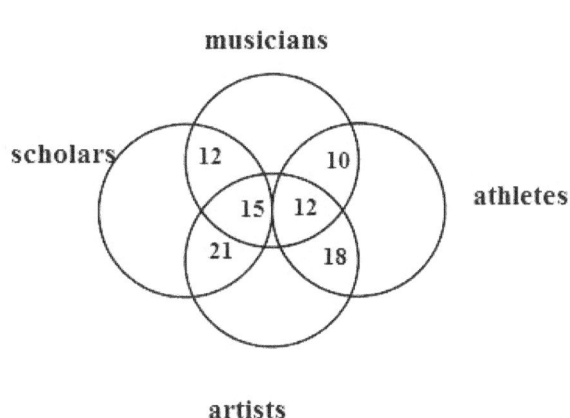

10. The above diagram summarizes a survey to evaluate the interests of a number of students. 100 students identified themselves as scholars, 100 as athletes, 100 as musicians and 100 as artists. A total of 285 students were surveyed. The overlapping circles indicate students who identified themselves as falling into more than one category. If one student were chosen randomly from the entire group of 285, which of the following is closest to the probability that that student identifies as an artist but not as any of the other three groups?

A) 0.12

B) 0.34

C) 0.66

D) 8.38

Explanation of 10:

There is a lot of information in this problem. Remember that much of that information may be completely unnecessary for solving the problem. Stay focused by keeping things concrete and simple.

Focus on what the problem is asking you to find. You are choosing one student at random. What is the probability that that student falls in the category of "artist" without overlapping with any other categories?

If the question stem is now clear to you, consider what you know about probability. A good intuitive strategy is to start with a very simple example. Say there is a group of ten friends and three of them have red hair. If you pick one friend at random, what is the probability that that person has red hair?

Is it clear from that example that you have three chances of picking a red-haired person out of a total of ten and that the probability can be expressed as $\frac{3}{10}$ or 0.3?

Hopefully a simple example can remind you that probability is the number of targets (the thing you are trying to pick – in the above example it was a red haired person) over the total number of people you are picking from.

Apply this to the current problem. How many people are there who are artists but are not identified with any other group? To do this you have to remember that there were 100 people who identified as artists. The graph shows you how many of those also identify with other areas. You can calculate how many are in the area of the graph that doesn't intersect with any other interest. It is 34.

You've got all the numbers you need. There are 285 total students. $\frac{34}{285}$ is the probability.

You'll notice that the answer choices are expressed as decimals. You could use your calculator here if you are very accurate with it. Let's see if there is an intuitive way to accurately get the right answer without the calculator.

First, lets' consider an approximation of the probability. $\frac{30}{300}$ is one tenth, so our answer is somewhere in that range. Consider choice A. It's close to one tenth, so it's in the running.

Choice B is about one third. A third of 285 is close to 100. That seems a bit high. Choice C, at about two thirds, is even higher. Choice D is out because our probability is a fraction less than one.

Without having to do any further calculation – which could lead to inaccuracy – we have a pretty clear answer.

If you are not quite confident in this approximation, you can try rewriting $\frac{34}{285}$, hopefully getting it to an equivalent fraction expressed over 100. Divide the top and bottom by 3. The result is $\frac{11}{95}$. Expressed as something over 100, you would have a number slightly bigger than 11 in the numerator. Choice A fits very nicely!

You finished Day 18! How did it go on these five questions?

Number of questions you got right on your own: _____

Types of problems or patterns you need more work on: _____

_____

How much new did you learn from these questions? ☐ Important tools! ☐ Some tools ☐ Not too much

Day 19, Questions 11-16

11. Jason held a bake sale and offered cookies for $2 each and cakes for $6 each. Between cookies and cakes, he sold a total of 15 items. At the end of the day he had received exactly $54 in sales. How many cakes did he sell?

A) 0

B) 6

C) 9

D) 36

Explanation of 11:

This problem can be solved with algebra but because algebra is very abstract, there is a big chance that you might make a mistake. Let's consider an intuitive approach that almost guarantees you a correct answer.

We can organize the information and the relationships using a chart with columns.

A	B	C	D	E
# cakes	# cookies (= 15-A)	$ from cakes (=6*A)	$ from cookies (=2*B)	Total income (C + D) Must be 54

Chapter 6. SAT®-style Calculator Questions with Explanations

In the chart, Column A is the number of cakes. We start with that because that's the number we're trying to find. If we test the answer choices, we can plug them into column A. Column B is the number of cookies but we already know that the total of cakes and cookies is 15, so if we subtract the number in column A from 15, we have column B.

Column C calculates the money received from selling the cakes, so that amount is $6 times the number in column A. Similarly, the money from cookies is $2 times the number in column B. Finally, to test whether the total money comes to $54, we add columns C and D.

If you do the process above, you are really thinking out all of the relationships in advance. This is a "big picture process", meaning that you are using wholistic, right brain thinking. Once you have completed this process, you'll begin doing actual calculations. Calculating is a "detail thinking" process. It is a concrete, left-brain process. Problem solving works best if you do your big picture thinking separately from your detail thinking. If you try to do them at the same time, it is easy to get confused.

In the example above you first focus on the Big Picture and when that is done you focus on the detail.

Once you've completed your chart of relationships, you are pretty much guaranteed a correct answer. Test the answer choices, starting with choice A, zero cakes.

A	B	C	D	E
# cakes	# cookies (= 15-A)	$ from cakes (=6*A)	$ from cookies (=2*B)	Total income (C + D) Must be 54
0	15	0	30	30
6	9	36	18	54

Choice B proves to be the correct answer. You don't have to test any further choices.

Let's step back and look at the algebra of this problem. It is a problem in two variables – say, a (for cakes) and o (for cookies.) We know that $6a + 2o = 54$, meaning 6 for each cake and 2 for each cookie equals $54.

Remember that an equation with two variables does **not** have a specific solution. There may be many combinations of a and c that would make the equation work. To have a unique solution for a system with two variables, you must have two different equations defining their relationship.

In this problem you do have a second equation: $a + o = 15$. The total number of cakes and cookies is 15. This is enough information to solve the problem.

$$6a + 2o = 54$$
$$a + o = 15$$

You could now use your intuitive tools for solving these two equations. If you double the bottom equation, the two *o* factors will be the same.

$$6a + 2o = 54$$
$$2a + 2o = 30$$

When you subtract the bottom from the top, you get

$$4a = 24, \text{ or } a = 6$$

Which method is best? That depends on how confident and how accurate you are in doing the algebra. If you sometimes get confused and mess up the algebra, it's better for you to use the more intuitive approach.

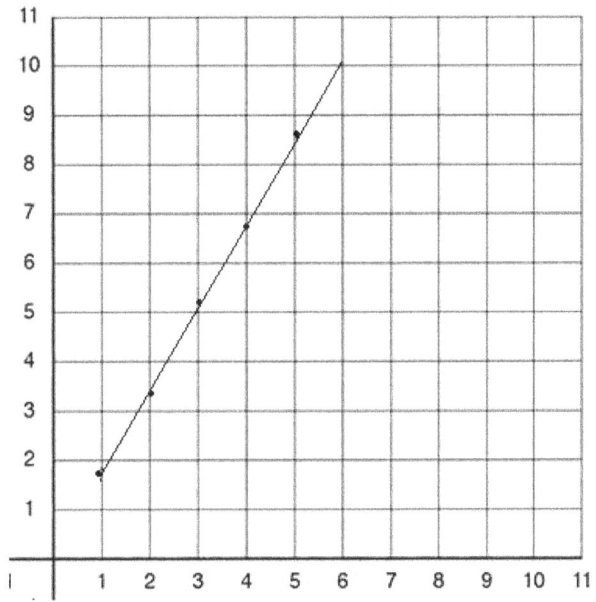

12. On the *xy* graph above, five points on the line segment are $(1, \frac{11}{6})$, $(2, \frac{21}{6})$, $(3, \frac{31}{6})$, $(4, \frac{41}{6})$ and $(5, \frac{51}{6})$. Which equation below correctly identifies the relationship between *x* and *y*?

A) $y = \frac{11}{6}x + 10$

B) $y = x^2 + \frac{11}{6}$

C) $y = \frac{10x}{6}$

D) $y = \frac{5}{3}x + \frac{1}{6}$

Explanation of 12:

This question asks you about the relationship between x and y. You have a number of intuitive clues that you can use. There is a hard way to do this problem and an easy way. We're going to do the hard way first because there is something important that you can learn from it and you may need to use the hard way on another question.

The hard way uses the generic formula for a line.

$$y = mx + b$$

where m represents the slope and b is a constant that gives you additional information about where the line is located. For a particular line there does not have to be a value for b. In other words, b can be zero.

Notice that most of the answer choices conform to the generic equation for a line. In choice C, b is zero. Choice B, however, involves a square of x and that would not result in a line, so choice B is out.

In choice A, the slope, m, is $\frac{11}{6}$. In choice C it is $\frac{10}{6}$. In choice D it is $\frac{5}{3}$, which is also $\frac{10}{6}$. If you can determine the slope, you may get the answer or at least be down to two possibilities.

Two find the slope, use the first two points. The run is 1 (going from $x = 1$ to $x = 2$) and the rise is $\frac{10}{6}$ (going from $\frac{11}{6}$ to $\frac{21}{6}$). The rise over the run is $\frac{10}{6}$ over 1, which matches choices C and D.

Carefully consider the difference between the two choices. The only significant difference is in the value of b. Let's test one or two known points to see which of the two answer choices they match.

Testing $(1, \frac{11}{6})$, we find that choice C gives:

$$\frac{11}{6} = \frac{10x}{6} = \frac{10*1}{6} = \frac{10}{6}$$

This is not a true statement. Choice C doesn't work. A quick glance at choice D tells you that it differs from choice C by adding on $\frac{1}{6}$. This gives us a y value of $\frac{11}{6}$, which is correct.

The way you just solved this requires that you understand $y = mx + b$. For some questions that approach will be the only way, or at least the best way, to get to the answer.

For this question the easier approach would have been to take one of the known points and see if it works in the answer choices. Any choice that doesn't work is out for sure. You may need to use a second or even third point to get to the answer.

This method completely avoids having to work with slopes or $y = mx + b$. Start with the first point, $(1, \frac{11}{6})$. Test it in choice A.

$$y = \frac{11}{6}x + 10$$

$$\frac{11}{6} = \frac{11}{6} * 1 + 10$$

$$\frac{11}{6} = \frac{11}{6} + 10$$

This doesn't work. Choice A is out. Try choice B.

$$y = x^2 + \frac{11}{6}$$

$$\frac{11}{6} = 1^2 + \frac{11}{6}$$

$$\frac{11}{6} = 1 + \frac{11}{6}$$

This also doesn't work. Try choice C.

$$y = \frac{10x}{6}$$

$$\frac{11}{6} = \frac{10 * 1}{6}$$

$$\frac{11}{6} = \frac{10}{6}$$

Choice C is also out. Test choice D to make sure it works. If you have made an error, choice D will also not work.

$$y = \frac{5}{3}x + \frac{1}{6}$$

$$\frac{11}{6} = (\frac{5}{3} * 1) + \frac{1}{6}$$

$$\frac{11}{6} = \frac{5}{3} + \frac{1}{6}$$

$$\frac{11}{6} = \frac{10}{6} + \frac{1}{6}$$

$$\frac{11}{6} = \frac{11}{6}$$

You've proven that choice D is the answer!

Questions 13 and 14 refer to the information below.

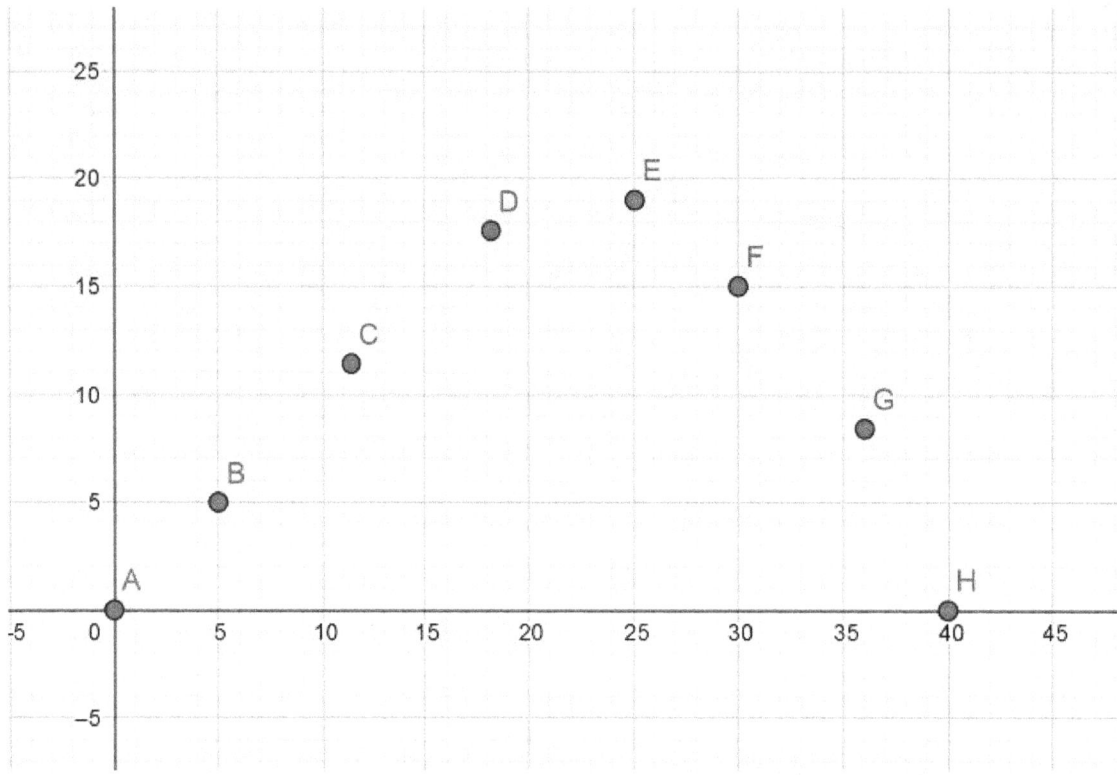

The chart above represents the growth of a yeast colony in a nutritive medium in a petri dish. The *y-axis* represents thousands of cells living at a given time. The *x*-axis represents hours. When all of the nutritive medium was consumed, the yeast stopped growing and existing yeast cells began to die.

13. Which of the following most closely represents the rate of growth during the slowest period of growth?

A) 1000 cells per hour

B) 300 cells per hour

C) 167 cells per minute

D) – 3000 (negative 3000) cells per hour

Explanation of 13:

As with all charts and graphs, your first task is to orient yourself carefully to the visual data. The *x*-axis represents time in hours – 5 hours, 10 hours, 15 hours, etc. The *y*-axis represents the number of yeast cells in thousands – 5000, 10,000, 15,000, etc – that area live at that moment. It's easy to forget later on that these are thousands, so make a mental note to be careful about that.

194   Chapter 6. SAT®-style Calculator Questions with Explanations

Continue to orient yourself to the data in the graph. The dots represent points in time and the number of yeast cells present at that time. You can make this information very concrete for yourself by looking at a few specific dots. Point A is at the origin. No time has passed and there are no yeast cells. Makes sense! At point B 5 minutes have passed and there are about 5000 yeast cells.

Next examine the trends in the graph. This is a very helpful step. What do you notice? The line goes up until about 25 minutes. What does that mean? The number of yeast cells is increasing. You may also notice that the rate of increase is pretty steady. The slope of the line is fairly constant. When does that change? After point D the slope of the line changes.

Notice that we're not talking about calculating the mathematical slope of the line (rise over run.) You can intuitively see the slope and then you can learn something about the relationships from what you intuitively observe.

After point D the line continues to go up a little until point E and then the line drops pretty sharply.

Now that you've oriented yourself to the data, let's come back to the question stem and orient to that. The question is asking for the rate of growth during the slowest period of growth. Let's make a road map.

Step 1. Find the slowest period of growth.
Step 2. Find the amount of growth during that period.
Step 3. Find the length in time of that period.
Step 4. Put #2 over #3. That gives us amount of growth over time, which equals the rate, amount of growth per unit to time.

Where is the slowest period of growth? Be careful! Someone might say "From G to H." However, during that period there is no growth. The setup tells us that when all the nutritive medium is consumed, growth stops. Can you see from the graph at what point growth stops? It would have to be point E, wouldn't it? After point E, there is no more growth. The number of yeast cells begins to decrease as the yeast cells die.

The last period of growth is from D to E. We've completed step 1 of our road map.

Step 2. How much growth was there in the period D to E? There's not a lot. Point D is at about 17,000. Point E is at about 19,000. The increase is about 2,000.

Step 3. What is the length in time of the period from D to E? D is at about 18 hours. E is at 25 hours. The difference is 7 hours.

Step 4. 2000 cells per 7 hours = $\frac{2000}{7}$ cells per hour ≈ 285 cells per hour

In Step 4 you don't necessarily need to use the calculator to divide 2000 by 7 if you notice that 7 would go into 2100 exactly 300 times. You could just approximate $\frac{2000}{7}$ as "a little under 300" or as "280 to 290".

Our next step is to review the answer choices. Choice A is about three times higher than our estimate. Choice B is very close to our estimate. It will be hard to beat that as an answer. Choice C is expressed in cells per minute, not cells per hour. Because there are 60 minutes in an hour, the hourly rate would be vastly higher than 167. Is it intuitively obvious that the number would be far beyond our estimate of 280 to 290? If it isn't intuitively obvious, then do the actual math to make sure! You don't need an exact number, just a rough approximation.

Choice B still looks like the best answer. As a final double check, consider choice D. It has a negative number. The setup tells us that growth stops when the medium is used up. Death of cells is not the same as a negative growth, so this is logically inconsistent. However, if you're not convinced of the logic, you can calculate the "negative growth" rate for the interval from G to H. It comes out to 2000 cells lost per hour. Choice D is much too high even if we are supposed to interpret cell death as negative growth.

Choice B is the correct answer.

---

14. Which of the following line segments most closely approximates the function $f(x) = x$?

A) $\overline{AH}$

B) $\overline{BH}$

C) $\overline{AD}$

D) $\overline{DE}$

Explanation of 14:

This question also refers to the yeast graph. If $f(x) = x$ seems a little intimidating, remember that it is exactly the same as saying $y = x$. The expression "f of x" stands for "a function of x" and tells us that there is a relationship between x and a number derived from x. The derived number is used as the y coordinate. The equation defines the relationship or function.

In this case the relationship is that the derived number (y) is the same as x. What does the line $y = x$ look like? Try out some points. You only need two points to know what the entire line looks like! If you graph (0,0), (5,5), (10,10), you can see that the line is at about a 45 degree angle. You can now look for parts of the graph that look like that. The whole first section of the line from A to D approximates that. Choice C is correct.

$\overline{AH}$ is a horizontal line. $\overline{BH}$ is nearly horizontal. $\overline{DE}$ is flattened out compared to $\overline{AD}$.

15. The table below shows the most popular pets in State X in order of the number of specimens in each category.

Fresh water fish	142 million
Cats	88.3 million
Dogs	74.8 million
Small animals	24.3 million
Birds	16 million
Horses	13.8 million
Reptiles	13.4 million
Saltwater fish	9.6 million
Total	**382.2 million**

Using the information in the table, what percent of the most popular pets are either reptiles or cats?

A) 27%

B) 30%

C) 33%

D) 45%

Explanation of 15:

A good reminder about percent is that it consists of a part of something over the whole of something, which creates a ratio, expressed in terms of a ratio over 100. It looks like this:

$$\frac{part}{whole} = \frac{?}{100}$$

That's your road map and it's a very powerful one! What is our "part" in this example? We have to combine reptiles and cats (never do this in real life!)

$$88.3 \text{ million cats} + 13.4 \text{ million reptiles} = 101.7 \text{ million}$$

If you are confident that you can do this addition accurately on your calculator, that's fine. If you're not, there are alternate ways to do this on paper. At the very least, you can use a hand calculation to double-check your calculator work. Here's how you might do this one:

$$88.3 + 13.4 = (85 + 3.3) + (10 + 3.4) = (85 + 3 + 0.3) + (10 + 3 + 0.4)$$

You can now rearrange these numbers.

Chapter 6. SAT®-style Calculator Questions with Explanations

$$(85 + 10) + (3 + 3) + (0.3 + 0.4)$$

$$95 + 6 + 0.7$$

And if 95 + 6 doesn't feel intuitively clear to you, break it down more!

$$95 + 5 + 1 + 0.7 = 101.7$$

Going back to our road map, notice that the 101.7 is the "part". Put it over the whole.

$$\frac{101.7 \, million}{382.2 \, million} = \frac{?}{100}$$

Notice that we can ignore the "million". Isn't it true that 3 million over 4 million can be reduced to 3 over 4? You could use the calculator to turn the fraction on the left into a decimal. Because it's easy to make mistakes on the calculator, let's look at an alternate method that you can use as a double check.

Approximation tells us that the numerator, 101.7, is very close to 100. We'll pretend that it is 100. The denominator is a little short of 400.

What would $\frac{100}{400}$ be when expressed as a percent? It's the same as $\frac{1}{4}$, which is 25%. Consider the answer choices. All the choices are larger than 25% and that is what we would expect from our estimation. Choice A is the closest and most likely the answer. However, you may need to test choice B to prove that it is too high.

This is the type of problem in which it would be very helpful to be highly accurate with your calculator. It would show you that the correct answer is 26.6%. If you aren't confident in your calculator skills, there are some intuitive strategies that you can use. If you are interested, consider the method below.

100%	382.2
10%	~38
5%	~19
1%	~3.8

In the above table we have identified some numbers that were very easy to do in one's head. The trick now is that if we want to know 20%, we can simply double the number for 10%. If we want to know 25%, we can add the number for 20% to the number for 5%.

Using this information, 20% would be about 76. Adding on the amount for 5% (19) gives us 95. (If you can't do that in your head, do 76 + (20 + -1) = (76 + 20) + -1 = 96 −1 = 95. The point of this method is to stick with simple calculations that you can do with perfect accuracy either in your head or on paper.

We now know that 25% is 95. We need to reach 27%, so need to know the figure for 2%. It is double the figure for 1% (3.8) and thus gives us 7.6. Adding that to 95 gives 102.6. This is extremely close to the 101.7 that we calculated.

You can now **prove** that choice B is too high. Adding 3% more, in order to go from 27% to 30%, would mean adding on 3.8 three times. This clearly takes us too far beyond the 101.7. You have proven that choice A is correct.

This problem brings up the issue of how much you should rely on your calculator. There are many intuitive ways to do even complex calculations without the calculator. If you are consistently able to do these kinds of calculations on the calculator without making errors, then go for it. If you tend to make errors on the calculator, learn how to do calculations intuitively so that you can double check your calculator work.

16. The relationship between y and x is defined as $y = kx + 3$. When $x = 8$, $y = 27$. For what value of x does $y = 54$?

A) 9
B) 16
C) 17
D) 165

Explanation of 16:

What do you need to do to solve this problem? They give you values for x and y. If you are not sure what to do next, put in what you know and see where it leads you!

$$y = kx + 3$$
$$27 = k*8 + 3$$
$$27 - 3 = k*8 + 3 - 3$$
$$24 = k*8$$
$$24 = 8k$$
$$3 = k$$

You have determined the value for k. The question asks you to find the value of x when y is 54. There is a trap here. 54 is twice 27, the original value of y when x was equal to 8. You might be tempted to say that if y doubles, then x should double. In fact, at this point you do not really know if that is true. Let's put the numbers that we know into an equation and find out for sure.

$$54 = 3x + 3$$

Solve for x. That is what the problem is asking you to find.

$$54 + -3 = 3x + 3 + -3$$
$$51 = 3x$$
$$17 = x$$

Choice C is the correct answer.

You finished Day 19! How did it go on these six questions?

Number of questions you got right on your own: _____

Types of problems or patterns you need more work on: _____

_____

How much new did you learn from these questions? ☐ Important tools! ☐ Some tools ☐ Not too much

Day 20, Questions 31-35

Instructions: For questions 31-35 there are no multiple-choice answers. You must solve the problem on your own. No answers are negative. Some problems have more than one correct answer. You only need to put down one correct answer. Your answer can be a fraction or decimal.

31. A school wants to build a water storage tank in order to be able to water the athletic field in case of a power outage. The tank will be a right circular cylinder measuring 16 meters in height. The school calculates that the tank needs to hold 40,000π cubic meters. What will be the <u>diameter</u> of the base of the tank in meters?

Explanation of 31:

Draw a picture. It almost always helps, even if you are just doodling. Orient yourself to the question. What is it asking you to do? You need to use the relationship between the dimensions of a right circular cylinder and the volume of the cylinder in order to work backwards from the volume to the diameter of the base of the tank.

Let's first review the relationships in a right circular cylinder. The important dimensions are the area of the base, the height, and the volume. To find the area of the base, we also need to know the diameter of the circle.

One good intuitive way to remember how to calculate the volume of the cylinder is the "pancake" model. First consider a flat circle lying on the ground. Suppose it has a radius of 5 feet. What is the area of the circle?

If you're not very clear on the area of a circle, here's a quick intuitive review of circles. There are two aspects of a circle that you are usually asked to calculate – the circumference and the area.

circumference   area

You can think of circumference as "walking around the outside of the circle." It is a distance measured in linear units (3 feet, 3 inches, 3 meters). The area is the amount of space in the inside of the circle. It must be measured in square units, just as you would measure the area of a room or of a rug (3 square feet, 3 square inches, 3 square meters.)

There is a formula for calculating circumference and a formula for calculating area. However, they look pretty similar and a lot of people have trouble remembering them and keeping them straight. Here's a way to understand it so that you'll never get confused again!

Both formulas include the same three elements: r, 2, and π (pi). The two formulas are:

$$2\pi r \text{ and } \pi r^2$$

Which one is which? The formula with the square in it is the formula for area (square units.)

Come back to our problem with the circle lying on the ground, with a radius of 5. What is the area of the circle?

$$\pi r^2 = \pi 5^2 = 25\pi \text{ square feet}$$

Now we're going to turn the circle into a pancake. Imagine that the circle is an extraordinarily large pancake that is exactly one foot thick. What was previously a two dimensional circle with an area is now a three dimensional shape with a volume. Because it is 1 unit high, the volume is 25π cubic feet.

The volume of a right three-dimensional figure is the area (in two dimensions) of its base times its height.

The above "pancake" has a volume of 25 square feet * 1 foot = 25 cubic feet. To create a cylinder, you can stack pancakes. You know the volume of each pancake. If you stack seven pancakes, the volume of the cylinder will be 7 * 25 cubic feet. The important thing is that each pancake is exactly one unit thick.

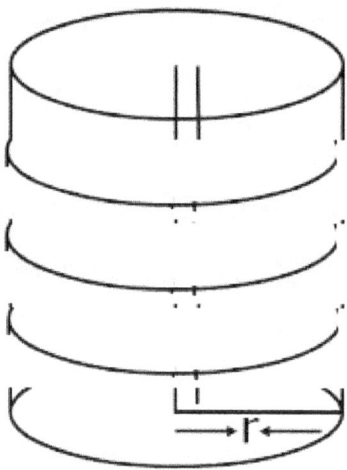

The cylinder above consists of four "pancakes", each of which is one unit high. Mathematically, the volume of the cylinder is described as:

$$\pi r^2 * h$$

where h is the height of the cylinder. You can understand now why this works. The πr^2 is the area of the circle that is flat on the ground. πr^2 times 1 unit of height gives you the volume of one "pancake". πr^2 time the height, h, gives the volume of the stack of pancakes.

Now that you've reviewed the relationships in a cylinder, let's get back to the problem. We are given h and the final volume:

$$\pi r^2 * 16 \text{ m} = 40{,}000\pi \text{ cu m}$$

You are asked to solve for the diameter of the base. Our formula does not contain diameter. However, diameter is simply twice the radius. Let's create a simple road map so that we don't forget to look for diameter.

> Step 1. Solve equation for r.
> Step 2. Multiply r times 2.

It may seem overly simple, but actually writing down your road map is a very powerful way to avoid common errors. Write it down on your scratch paper. Don't try to keep it in your head!

Solving for r:

$$\pi r^2 * 16 \text{ m} = 40{,}000\pi \text{ cu m}$$

$$r^2 * 16 \text{ m} = 40{,}000 \text{ (divide both sides by } \pi\text{)}$$

$$r^2 = \frac{40000}{16} \text{ cu m (divide both sides by 16)}$$

$$r^2 = \frac{20000}{8} = \frac{10000}{4} = \frac{5000}{2} = 2500$$

$$r^2 = 2500 = 25 * 100 = 5 * 5 * 10 * 10 = (5*10)(5*10) = (5*10)^2 = 50^2$$

$$r = 50$$

Check the road map. Multiply r times 2. The diameter must be 100 meters.

Notice that in the above calculations we avoided the calculator by doing a series of very easy simplifications. Instead of dividing out $\frac{40000}{16}$, we simply reduced the top and bottom by two a number of times until we reached 2500.

Then, instead of using a calculator to figure the square root of 2500, we divided it into factors. This only works, of course, if the number **can** be divided into factors that yield a nice, clear square root. However, it's worth giving it a try! It often allows you to approximate the square root.

Year	Low temp	Year	Low temp
1990	46	1996	38
1991	35	1997	41
1992	32	1998	45
1993	43	1999	33
1994	41	2000	36
1995	37	2001	40

32. The table above shows the low temperature for the month of June in a particular city over a period of 12 years. Based on this table, what is the mean low temperature for June for this city during these 12 years? (Round your answer to the nearest tenth.)

Explanation of 32:

The question is asking you to find the mean of the twelve temperature values given in the chart. The terms "mean" and "median" are often confused. "Mean" **means** average. They are the same thing. Average "means" mean. The median is something different.

The usual way to calculate the mean (which **means** average) is to add all the numbers and divide by how many total numbers there are – in this case, 12. If you are highly accurate with adding **and** dividing on the calculator, this can be quick. Because this question isn't multiple choice, you will have to calculate an exact number. The calculator may be a good tool for this. However, there are intuitive tools for adding a long series of numbers accurately on paper. (If you use the calculator, double-check your results or do the calculation a second time.)

To add intuitively, divide each number into its "easy" part and the remainder. The easy part ends in 0 or 5.

Original	Easy part	Remainder
46	45	1
35	35	0
32	30	2
43	40	3
41	40	1
37	35	2
38	35	3
41	40	1
45	45	0
33	30	3
36	35	1
40	40	0

You can regroup these numbers for easy calculation. For example, two 35's is 70. Add both columns and combine them. To divide by 12, which is 2*2*3, you can divide by 2 twice and then divide by 3. The correct answer is 38.9.

Questions 33 and 34 refer to the following information.

Language	Italian	English	Russian
Avg # of words	550	865	720
Average speed (words per minute	80	100	60

The table above summarizes Dale's word processing skills for short stories in three languages. The average number of words refers to the number of words in an average short story that Dale processes, which is different for each language. The average speed refers to the average number of words that Dale can type per minute in each language.

33. If Dale word processes a project consisting of one story of average length in Italian and Russian each and two stories of average length in English, and the project takes Dale exactly one hour, what is Dale's average speed in words per minute?

Explanation of 33:

This is a rate question. You can think of a rate as a certain amount of something over a certain amount of time, for example miles per hour or words per minute. It is helpful to remember with rates that you usually cannot simply add two rates together. Why not?

Consider an example of a race car driver who starts out in car A going 100 miles per hour and then part way through the race switches to car B going 120 miles per hour. What is the driver's average rate (speed)? It is **not** necessarily 110 miles per hour. If the driver drove car A for a quarter mile and car B for two miles, you can see that the average rate is not just the average of the two rates.

To determine a rate, we need to find the total "amount" and the total "time" and use those numbers to create a new rate.

In this problem we will need to find the total number of words that Dale types and the total time it takes Dale to type them. We will use the given rates to figure words and times for each language but we will then have to add up all the words and then add up all the times and calculate words per minute.

There are a lot of steps to this problem so this is a good time to create a road map and write it down on paper.

Step 1. Add the number of words in an average Italian story to the number in an average Russian story. To that, add double the number of words in an average English story. This gives total words.

Chapter 6. SAT®-style Calculator Questions with Explanations

Step 2. Calculate the time that each language will take. The setup tells us that the project will take one hour, which must be converted to 60 minutes because we need words per minute.

Step 3. Divide the answer from 1 by the answer from 2. This gives words per minute.

Step 1. We know the number of words in the average story of each language so this is relatively straightforward. 550 + 720 + 865 + 865 (two stories in English.) = 3000.

This addition can be done intuitively, rather than by calculator or hand, by recombining the elements.

$$550 + 720 + 865 + 865$$
$$500 + 50 + 700 + 20 + 800 + 15 + 50 + 800 + 15 + 50$$
$$500 + 700 + 800 + 800 + 50 + 50 + 50 + 15 + 15 + 20$$

Combine elements that are easy to add:
$$(700 + 800) + 500 + 800 + (50 + 50) + 50 + (15 + 15) + 20$$
$$(1500) + 500 + 800 + (100) + 50 + (30) + 20$$

Regroup:
$$(1500 + 500) + (800 + 100) + (50 + 50)$$
$$2000 + 900 + 100 = 3000$$

If that looks a little complicated, you might find it easier if you organize it in a slightly different way.

Number	Easy part	Left over	Grand total
550	500	50	
720	700	20	
865	800	65	
865	800	65	
Totals	2800	200	3000

You can break this process down even more depending on how comfortable you are with adding numbers like 500, 700, 800 and 800 in your head.

Number	Easy part	Left over 1	Left over 2	Left over 3	Grand total
550	500		50		
720	500	200		20	
865	500	300	50	15	
865	500	300	50	15	
	2000	800	150	50	3000

We are still at step 1, having calculated the total number of words at 3000. For step 2 we know that the time is 60 minutes. In step 3 we divide 3000 words by 60 minutes. How do we know that we have to divide, rather than, say, multiply, or divide 60 by 3000?

A very powerful way to keep this straight is to use "factoring by units." We need to end up with units of words per minute. Consider the data that we have:

$$3000 \text{ words} \quad 60 \text{ minutes}$$

To get $\dfrac{words}{min}$ we simply have to put words in the numerator and minutes in the denominator:

$$\dfrac{3000 \, words}{60 \, min} = 50 \text{ words per minute}$$

34. On Tuesday, Dale completes a project consisting of one average length story in each language, typed at the average speed for each language. On Wednesday Dale types another project consisting of one average length story in each language. Dale types at the average speed for the Italian and English stories. If Dale's typing on Wednesday for the Russian story is 20% slower than the average speed, how many more minutes will it take to complete the Wednesday project compared to the Tuesday project?

Explanation of 34:

Take a minute to orient yourself carefully to this setup. You are comparing two scenarios. Each scenario has three parts but in the second scenario, only one part has changed. Wouldn't it be true that you only have to deal with the part that changed?

Using intuitive tools, let's look at that graphically to see if it clarifies the situation.

Scenario	Italian	English	Russian	Total time
Tuesday	550 words at 80 wpm	865 at 100 wpm	720 at 60 wpm	I + E + R
Wednesday	Italian: 550 words at 80 wpm	English: 865 at 100 wpm	Russian: 720 at 20% slower than 60 wpm	I + E + ?

In the Total time column "I" stands for the time it will take to type the Italian story, "E" stands for the time for English, and "R" stands for the time for Russian at the average rate.

The only difference between the time for scenario 1 and scenario 2 is the Russian time. Does the table help clarify that?

Use intuitive common sense to orient yourself to what will happen when Dale types the Russian story more slowly on Wednesday. It will take more time. The question stem tells us that but by making sure you are oriented to this, you may avoid some calculation errors.

Let's create (and write down!) a road map.

Step 1. Calculate the time for Russian on Tuesday.
Step 2. Calculate what number of words per minute would be 20% slower than 60.
    a) find 20% of 60
    b) subtract "a" from 60
Step 3. Calculate the time for Russian on Wednesday, using the answer to step 2.
Step 4. Subtract step 2 from step 3.

Wednesday time	
Minus Tuesday time	
Difference	

The table above is part of the road map. You will use it to fill in the information as you get it.

Step 1. Use factoring units to find the number of minutes for Russian on Tuesday. We want minutes, so the unit "minutes" has to be on top.

$$\frac{1 \text{ min}}{60 \text{ words}} * \frac{720 \text{ words}}{1} = \frac{720}{60} = \frac{72}{6} = \frac{36}{3} = 12 \text{ minutes}$$

Notice how we could reduce the original fraction one small step at a time by dividing in half several times and then dividing by 3. This avoids possible errors in using the calculator or trying to divide 60 directly into 720.

Enter your result in the road map table.

Wednesday time	
Minus Tuesday time	12
Difference	

Step 2. What is 20% slower than 60 words per minute? This looks deceptively simple. It's not. First, orient yourself to the fact that the answer must be lower than 60. That might seem obvious but it's amazing how simple things like that can get lost during the test and someone comes up with a number like 72!

There are a number of ways that you can use to understand what 20% lower is. One way to think of it is that you need to find 20% of 60 and then subtract that amount from 60. You can also think of that as finding 80% of 60.

Another intuitive way to organize this is visually.

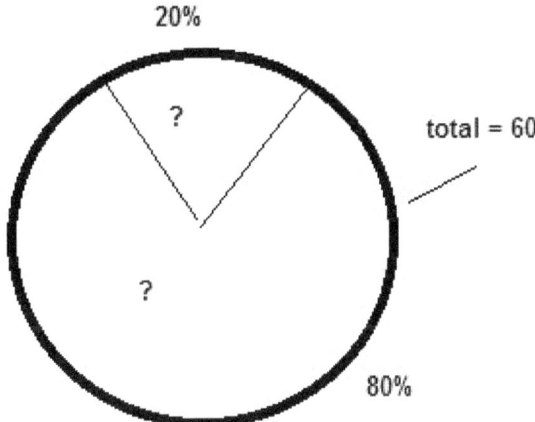

Does this help you understand more accurately what removing 20% of 60 would be?

Using any of these methods, we can calculate that 20% of 60 is 12. You can use your calculator to get this if you are extremely confident in your accuracy with the calculator. Otherwise, if you recognize that 20% is $\frac{20}{100}$ and is equivalent to $\frac{1}{5}$, you can divide 60 by 5. If you aren't comfortable with your accuracy with that, here is a way to use a table to organize the information.

60	10%	6
	20%	6 + 6 = 12
	80%	12*4 or 6*8

It's easy to find 10% of a number. Once you have that, you can simply double it to get 20%.

Following the road map, we subtract 12 from 60 and get 48 words per minute as the new rate for Wednesday's Russian story. Notice that this is 80% of 60. You can double-check your answer in the table above. 80% is 4 times 20% or 8 times 10%.

Remember that **many** errors on the test are simply due to not double-checking. You knew what to do but slipped up somewhere. A major way to increase your score is to learn to double check using intuitive tools!

Step 3. Using the new rate for Wednesday's Russian story, we can calculate how much time it will take.

$$\frac{1 \min}{48 words} * \frac{720 words}{1} = \frac{720}{48} = \frac{360}{24} = \frac{180}{12} = \frac{90}{6} = \frac{30}{2} = 15 \text{ minutes}$$

Wednesday time	15
Minus Tuesday time	12
Difference	3

Step 4. Your road map table has now made this problem foolproof. The answer is 3 minutes.

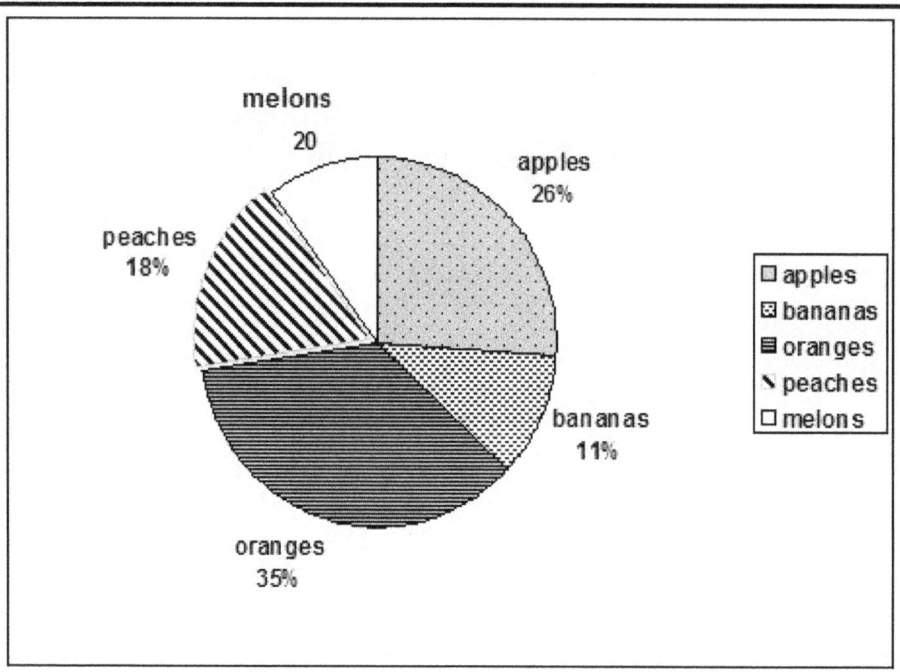

35. May received a box consisting of individual pieces of various fruits as a gift. The above pie chart shows the percentage of apples, bananas, oranges, and peaches that she received. The remaining 20 pieces of fruit were melons. How many more oranges did May receive than bananas?

Explanation of 35:

As always with a graphic, orient yourself carefully to it before starting. Many errors are due to the person simply not having really understood the graphic information. In this case we have a "whole", that is the entire contents of the gift box, broken up into a number of parts. Look carefully at what you know about each part. There is a trap here!

For four of the fruits we know the percentage of the whole that the fruit constitutes. For example, peaches are 18% of whole. However, for melons we know only an exact number of fruits. There are 20 melons, **not** 20%.

Now orient yourself to the question stem. You are asked to find how many more oranges there are than bananas. To avoid making a careless error, it helps to notice that oranges constitute 35% and bananas 11%, so there are clearly more oranges than bananas.

The most direct way to the answer is to find out the number of oranges and the number of bananas, right? What do we know? Only percents. Does the problem tell you how many total fruits there are? No.

The questions in the previous paragraph area a great illustration of the KING approach to solving problems. In the KING approach you ask yourself questions in order to problem solve.

K = Know. What do I already know?

I = Infer. Given what I know, what can I infer from that? What else must be true?
N – Need. What information would I need in order to get the answer?
G – Get. How can I get the information I need?

We know percent of the whole for oranges and bananas, as well as two of the other fruits, but we don't know how many there are. We know how many melons there are. Do we know what percent the melons constitute? Not directly, but if you look at the graph, the other fruits account for (add them up) 90% of the whole. We can infer that 20 melons is the remaining 10%. And now we can infer that if 10% is 20, then 100% is 200. There are 200 total fruits in the whole.

The oranges constitute 35% of 200 and the bananas constitute 11% of 200. Of course 200 is twice 100. 35% of 100 is 35, so 35% of 200 is 70. Similarly, you can easily calculate that there are 22 bananas. The difference between 70 and 22 is 48. The answer is that there are 48 more oranges than bananas.

An alternate way to get the answer that can serve as a double check is to calculate the difference in percent between oranges and bananas. 35 – 11= 24. You can multiply the 24% difference times the total fruits (200) and also come up with 48 more oranges than bananas.

Often there is no one right, best way to solve a problem. It's helpful to be familiar with a number of alternative strategies. At any given moment one strategy might just not make sense to you and an alternative might save the day!

You finished Day 20! How did it go on these five questions?

Number of questions you got right on your own: _____

Types of problems or patterns you need more work on: _____

_____

How much new did you learn from these questions? ☐ Important tools! ☐ Some tools ☐ Not too much

Congratulations on getting through all of the questions!

You can put this book aside for a week or two and then go through it again to test yourself. See how much you remember. Look for areas that are still challenging for you. As you work on practice questions from other books, try to use your intuitive strategies whenever you can. When you find problems that are giving you trouble, use the Index on the following pages to review strategies for specific types of patterns. Good luck and enjoy your powerful new intuitive strategies!

Index of Problem Types

Under each heading, the entries, such as ACT® 1 or SAT® Calculator 2, refer to a chapter and a problem number.

Absolute value
ACT® 1
ACT® 22
ACT® 50
SAT® Calculator 2

Angles, complementary
ACT® 48
SAT® NonCalculator 17
SAT® Calculator 1

Area
ACT® 21
SAT® Calculator 4

Average/mean
ACT® 7,
ACT® 47

Census
ACT® 19

Circle
ACT® 18
ACT® 40
SAT® Calculator 10

Circle and triangle
ACT® 18

Circle, arc
ACT® 18

Circle diagrams
ACT® 40
SAT® Calculator 10

Constant

SAT® NonCalculator 7
SAT® Calculator 6
SAT® Calculator 9
SAT® Calculator 16

Cosine
    ACT® 17
    ACT® 46
    SAT® NonCalculator 20

Diameter of base
    SAT® Calculator 31

Digits
    ACT® 29

Equivalent expression/solution
    ACT® 11
    ACT® 12
    ACT® 14
    ACT® 44
    ACT® 51
    SAT® NonCalculator 2
    SAT® NonCalculator 11

Exponent
    ACT® 14
    ACT® 38
    SAT® NonCalculator 9
    SAT® NonCalculator 19

Exponent, fractional
    ACT® 38
    SAT® NonCalculator 9

Exponent, root of
    SAT® NonCalculator 19

Expression must be negative
    ACT® 58

$f(x)$
    ACT® 2
    ACT® 17
    ACT® 28

    ACT® 50
    ACT® 54
    ACT® 57
    SAT® NonCalculator 3
    SAT® Calculator 14

Factorial
    ACT® 10

Find a constant
    SAT® Calculator 9
    SAT® Calculator 16

Find the same solution
    ACT® 1

Find x
    ACT® 22
    ACT® 52
    ACT® 56
    SAT® NonCalculator 6
    SAT® NonCalculator 16
    SAT® NonCalculator 18

Find x and find y
    ACT® 52

FOIL
    ACT® 14

Formula for data set
    ACT® 55

Fraction
    ACT® 38
    SAT® NonCalculator 9
    SAT® NonCalculator 18
    SAT® Calculator 2

Fraction, undefined
    SAT® Calculator 2

Graph
    SAT® NonCalculator 15
    SAT® Calculator 5

SAT® Calculator 13

Imaginary numbers
    SAT® NonCalculator 14

Inequality
    ACT® 11
    ACT® 41
    ACT® 44

Infinite solutions
    ACT® 35

Inverse tangent
    ACT® 27

Line
    ACT® 48
    SAT® NonCalculator 13
    SAT® NonCalculator 15
    SAT® Calculator 1
    SAT® Calculator 12

Line, graph of
    SAT® NonCalculator 15

Lines, parallel
    ACT® 48
    SAT® Calculator 1

Log
    ACT® 31

$m^2 - n^2$
    SAT® NonCalculator 11

Match point to equation
    SAT® NonCalculator 12
    SAT® Calculator 12

Matrix
    ACT® 45

Mean/average
    ACT® 7

ACT® 43
ACT® 47
SAT® Calculator 32

Mean and median
ACT® 7

Median
ACT® 4
ACT® 7

Midpoint
ACT® 13
ACT® 23

Multiplication
ACT® 23

Number line
ACT® 16
ACT® 41

Percent
ACT® 33
ACT® 36
ACT® 39
ACT® 59
SAT® Calculator 15
SAT® Calculator 35

Perimeter
ACT® 21

Period
ACT® 2

Pie chart
ACT® 3
ACT® 34
SAT® Calculator 35

Polynomials, addition
SAT® Calculator 8

Probability

    ACT® 10
    ACT® 25
    ACT® 26
    ACT® 34
    ACT® 42
    SAT® Calculator 10

Pythagorean theorem
    ACT® 42
    SAT® NonCalculator 20

Quadratic expression
    ACT® 22
    ACT® 28
    ACT® 51
    ACT® 54
    ACT® 56
    SAT® NonCalculator 2
    SAT® NonCalculator 5
    SAT® NonCalculator 8
    SAT® NonCalculator 11
    SAT® Calculator 8

Random versus nonrandom
    ACT® 19

Rate
    SAT® NonCalculator 1
    SAT® NonCalculator 4
    SAT® Calculator 33
    SAT® Calculator 34

Ratio
    ACT® 30
    ACT® 37
    SAT® Calculator 4
    SAT® Calculator 10
    SAT® Calculator 35

Ratio versus actual number
    SAT® Calculator 35

Rectangle
    ACT® 21

Rectangular prism
    ACT® 9

Relationship between x and y
    SAT® Calculator 3

Relative position
    ACT® 16
    ACT® 53

Right cylinder
    ACT® 6
    SAT® NonCalculator 10
    SAT® Calculator 31

Right triangle
    ACT® 24
    ACT® 27
    ACT® 46
    SAT® NonCalculator 20

Root
    ACT® 54
    SAT® NonCalculator 9
    SAT® NonCalculator 18
    SAT® NonCalculator 19

Scatterplot
    SAT® Calculator 9

Sets
    ACT® 3
    ACT® 40
    SAT® Calculator 10

Sets, overlapping
    ACT® 40
    SAT® Calculator 10

$\sin^2 + \cos^2$
    ACT® 17

Sine
    ACT® 2

ACT® 17
ACT® 27
ACT® 46
SAT® NonCalculator 20

Slope
ACT® 13
ACT® 15
ACT® 23
SAT® Calculator 12

Square quadrilateral
ACT® 8

Square root
ACT® 54
SAT® NonCalculator 18

Squaring
SAT® Calculator 7

Surface area
ACT® 6

System of equations
ACT® 35
SAT® NonCalculator 5
SAT® NonCalculator 16
SAT® Calculator 8

Table
SAT® Calculator 15

$\tan^{-1}$
ACT® 27

Tangent, inverse
ACT® 27

Trapezoid
ACT® 32

Triangles
SAT® NonCalculator 17

Triangles, 3:4:5
    ACT® 24
    ACT® 46

Triangles, isosceles
    SAT® NonCalculator 17

Two variables
    ACT® 36
    SAT® NonCalculator 13
    SAT® Calculator 11
    SAT® Calculator 16

Volume
    ACT® 9
    ACT® 60
    SAT® NonCalculator 10

Volume, compare two
    SAT® NonCalculator 10

When will two events coincide
    ACT® 5
    ACT® 20

x,y coordinates
    ACT® 13
    ACT® 15
    ACT® 17
    ACT® 42
    SAT® NonCalculator 12
    SAT® NonCalculator 15
    SAT® Calculator 3
    SAT® Calculator 12
    SAT® Calculator 13

Bonus Material

Below is an excerpt from my Young Adult novel *Annie Gomez and the Gigantic Foot of Doom.* I hope you enjoy it as a reward for your efforts! The entire novel is available on Amazon.

Reviews of Annie Gomez and the Gigantic Foot of Doom:

"I didn't really get it."
Ed Ferguson, gym teacher

"I like to read in the bathroom and this was as good as anything. Well, almost anything."
Francine Botachek, retired cheerleader

"Brilliant story of a young magician whose parents were killed by the greatest evil of all time. Despite tremendous odds, he triumphs in the end, but not after an extended series of adventures with his chums. Oh, wait. You mean this little book over here?"
Denver Postit

"I like started at the beginning of it and then I like sorta forgot where I was and sorta slid into the middle, which was cool. It was cool, you know what I mean, and then I had some Ritz crackers and I'm kinda like wondering what happened to my pants. Anyway…"
Can I get back to ya on the name thing?

"It was ok but it's not the NY Times."
The NY Times

"The children in this book are not real! Don't ever do what they do! Problems must be solved by adults. And for goodness sake put your pants on."
Your dad

"If Jay Cutts is not yet [in] an institution, he should be, and soon."
Better Homes and Gargoyles

Preface

I think it's safe to say that life is unfair, adults are the cause, and young people are the solution. It's safe to say because I'm hiding behind this book and no one can throw anything at me.

I've been a teenager now for over 50 years, so trust me when I say that it's intense. One minute I'm ecstatic about some cool thing that I've figured out how to do[1] and the next minute I'm a trembling idiot because I have to talk to someone who is actually looking at me.

Please forgive me if I'm honest enough to say that I'm smarter than almost everyone I know and I don't get why they don't like me. Why can't they just accept that they're inferior? This is one thing I haven't figured out yet.

As Annie discovers, it's great to have friends who get you and who are smart in their own ways (though probably not as smart as me). So I'm grateful for our family's dog, Baxter, who is highly creative, very hands-on, and loves to have his butt scratched. How can you not love a friend like that? Baxter knows the words "walkies" and "mailbox", which puts him ahead of a lot of people I know. I'm also grateful for my human friends, whose names I've forgotten for the moment.

I think you're going to like Annie and her friends. Annie is that girl in school who's taller than you, cute, smarter than you (but not smarter than me), says funny things without even knowing they're funny, and who you'd like to hang out with but who you're afraid to talk to. Take my advice. It's like a rattlesnake. She's just as afraid of you as you are of her. So go ahead and say something to her. You'll both probably make idiots of yourselves but at least you will have tried. Just like rattlesnakes.

Annie has put together a sort of club. She calls it a "coterie". It includes a handful of her best friends. Actually her only friends. But, hey, as with rattlesnakes, quality is more important than quantity[2]. And her friends have really amazing quality.

So sit back, relax, and find out how Annie deals with the fact that two different sets of aliens are each claiming that the other one is trying to destroy humanity. How will she figure out which is which? How will she stop an entire alien race with powers far beyond those of mere mortals (probably even me)? What will she do when her last resort fails? Not to mention the question of whether true love is possible between beings from two different dimensions.

I think that in this book I've finally proven that the world should be run by teenagers, for a number of reasons, not the least of which involve pizza and not having to wear pants. Enjoy life! It only gets better!

Jay Cutts
Albuquerque, NM
January 14, 2015

---

[1] For example making dog food out of breakfast cereal and vice versa.

[2] I'm not sure why. Perhaps one good snake is better than a handful of mediocre snakes, but then who wants a handful of snakes, anyway?

Chapter 1

Truth, Justice, and the Goatery

When a 10th grade girl who has flunked every quiz and test for the entire semester aces her biology final with a perfect score, there is only one explanation. What bothered Dr. Tripledoor, the biology teacher, more than the score itself, though, was the answer that Annie Gomez had put down for the extra credit essay question: *What in your opinion most makes biology relevant to our lives?* Instead of the ever popular *I love biology because it lets us eat and sleep* or (giggle, giggle) *Without biology there would be no reason for Saturday night dates*, Annie had written:

The human race faces the very real and immediate danger of total extinction NOW.

She had also decorated each end of her sentence with a very neat but not biologically accurate picture of a flower, colored with pink highlighter.

The reason this answer caused Dr. Tripledoor undue anxiety was that he feared she was right. There was only one thing that would explain a 10th grade girl flunking every test and quiz and then achieving a perfect score *and* writing that essay. Alien intervention.

Dr. Tripledoor stroked his small, graying goatee and stared into space. He wasn't musing or contemplating or wondering. He was purposely staring into space because that, he knew, was where the danger, and just possibly the solution, lay. And, truth be told (though Dr. Tripledoor may not have liked it to be told) he was gripped by a cold, hopeless fear. Compared to the immensity of space, he suddenly felt infinitely tiny. Compared to space, he was, in fact, infinitely tiny[3], but being a biologist and not an astronomer, he had never considered the fact before. To make things worse, he suddenly realized that, as it was five o'clock on Friday, he was probably the only person left in the school building. A shiver ran up his spine. He jumped up and began collecting the papers he still had to grade, along with his lunch bag, walking stick, sunglasses, and a freeze-dried scorpion that he planned to add to his "terrarium of death" (his favorite hobby) over the weekend.

Before he left the room, he wrote a small note to himself and stuck it on the middle of his desk. "Monday. Warn Annie." Then, keeping his eyes down, he scurried (in a shuffling sort of way) out of the building, into his car, and away.

Annie Gomez was not the type of person who usually needed warning. She was also not the type of person who usually flunked quizzes and tests. She was, even by her own acknowledgement (though she never said so out loud) the brightest person in Highbotham High School. Something had to be terribly wrong for her to mess up so badly. Something far beyond the normal terribly wrong things that she had recently become aware of. Injustice, for example. It had been just at the beginning of the school year – only nine months ago now – that she had noticed that not only was injustice rampant but that its opposite – the supposedly noble justice – hardly seemed to exist at all.

---

[3] 66.4 liters compared to 2 x 10 to the 33rd power cubic light-years. No one knows how many liters fit in a cubic light-year. The best estimate so far is "a heck of a lot, so don't even *think* about trying to fill a cubic light-year with liters!"

How just was it, to take one random example, that she, the smartest kid in the school and the tallest girl in 10th grade, hardly had any friends? That most of the other students in her class couldn't find anything even faintly interesting to talk about? That the girls were all obsessed with hair and makeup, whereas she was obsessed with justice, hair, and makeup? That boys could barely talk to her at all, except for cracking jokes that would strike a third grader as unsophisticated?[4]

Clearly, injustice was rampant and the main victim of it was her. But not only her! There were others as reviled and denigrated[5]. It had been her task to find these people and protect them. That was how Annie's Coterie[6] had come into being at the beginning of the school year. The AC (as it was referred to by the members of the AC, as opposed to the Goatery, as it was referred to by those who were not members of the Goatery) currently comprised six fellow miscreants. Annie had carefully chosen students who were outsiders, who were radically different, and whom the mindless pack of normal kids instinctively shunned. And of course her members all adored her. Who wouldn't?

Her first recruit had been Andy Kanayurak. Andy's father was Inuit (Eskimo, to the uninitiated). Andy's mother was African-American. Nobody knew what to make of Andy. The round, cheery cheeks and almond-shaped eyes he inherited from his father twinkled like Arctic snow. The chocolate skin he inherited from his mother spoke of the African sun. His father's genes had relaxed his hair just enough to make his abundant Afro cascade like a fountain. He seemed to transcend race and that scared a lot of kids. It's also what made Andy incredibly cool. He would tell people, "Hey, race is a non-issue. If you went far enough back and figured out who your ancestors really were, everybody would seem like your cousin." Andy had a fantastic sense of humor about identity. If a cop hassled him, he'd say, "Is this because I'm an Eskimo?" which usually left the officer with his mouth hanging open.

Andy was the second smartest person at Highbotham. According to Annie, there were a number of second smartest people but most of them were smart in one particular area. Andy was smart in everything. He was even a good cook. His best grades were in math[7]. However, his real passion was theatre. He loved becoming a new character and bringing that character to life. His portrayal of Anne Frank (in drag and with serious amounts of makeup) had brought tears to the eyes of, well, none of the students, since it is not at all cool to cry in high school, but to most of the faculty and parents in the audience. Even Keri Jenkins, reporter for the school newspaper had admitted:

---

[4] In third grade, Annie knew what *unsophisticated* meant. It had been her personal word of the day for October 16. She could also spell it and give its Latin derivation. None of the teachers knew whether she was correct.

[5] Words of the day for April 12 in 1st grade and December 7 in 6th grade, respectively.

[6] *Coterie: Close-knit group of people with a similar purpose, often exclusive. From Middle French, meaning people sharing the same cot.* Word of the day, November 9, 9th grade. Probably the Middle French would have just called Annie's Coterie a gang. But then the Middle French didn't get out much, being stuck between the Outside French and the Top and Bottom French.

[7] His 8th grade teacher was forced to give him an A+ even though his test average was only A- because Andy had developed a system of equations for calculating the amount of time contained in a black hole.

Andy Kanayurak's performance as a Black Eskimo Anne Frank was the most unusual thing that this reporter has ever seen, and I've seen Sharon Anderson in a bikini (no offense, Sharon).

The second member that Annie had recruited for the AC was Justin Larson. She had chosen him out of pity. Justin looked like a little 5th grader who had gotten lost in the high school. With his shaggy blond hair hanging over his ears, he also looked a good bit like a golden retriever. Everybody liked him but in a head-patting sort of way. This might have been fine if Justin actually were a 5th grader and/or[8] a golden retriever, but Justin was a 10th grader with college credits in biochemistry, who toured internationally every summer with a world class jazz band. He played trombone. Annie had been the only one who had seen through his puppyesque veneer and he was eternally grateful to her (but was careful not to wag his tail.)

Annie's third recruit had transferred to Highbotham High at the beginning of the school year and by October he was so miserable that a cloud of doom hung over him. Dressed in black, a black hoodie pulled over his head whenever he could get away with it, and long, stringy black hair hiding his face, he hunched his tall, skinny body around the halls, avoiding all human contact. Johnny Dinicu had moved a lot of times and every time he started a new school, it was torture. All the other kids knew each other. He felt like a zit on the face of humanity.

But this time it had been worse. Somebody had found out about his family almost immediately. It couldn't have been more than a week into school when he was sitting at his desk quietly waiting for history class to start and some big muscley guy – Greg something or other – had shouted out, "Hey, Gypsy! Play the tambourine for us." That got a huge laugh for about 2 and a half seconds until Ms. Crimmins looked up and gave Greg a stare that made him choke half to death on his own laughter. You could literally hear the spit trying to go down his windpipe and drown him.

Ms. Crimmins' impromptu lesson on the history of the Romani people that followed just made things worse. Now everybody stared at him like he was a freak in the side show. "No, I'm not a Gypsy" he had said over and over. "There's no such thing as Gypsies. It's a made up idea by people who don't know who we are." His mother said that sometimes. She was a Romani rights activist. But even she knew there were times to keep your mouth shut about who you were.

Johnny had sat at a table by himself at lunch that day wishing he could go back to being an invisible nobody. Somebody with long, wavy golden-brown hair over her shoulders who smelled nice and was as tall as he was sat down next to him. Right next to him. Not across the table. Not a few feet away. They had the following weird conversation.

---

[8] There is one recorded case of a golden retriever being enrolled in the school system of a huge metropolitan area, under a false social security number, and making it to the 6th grade before being found out. The owners claimed that the dog enrolled himself but the dog had snapshots of the owners filling out the enrollment forms online. The dog went on to become a successful attorney by taking home study courses.

"*Droboj tu*[9]," Annie had said.

"*Najis*," he had mumbled, focusing on his tater tots and trying not to pay any attention to her. Then it hit him. "What the … What did you say to me?"

"You heard me, dude," she said. "Don't worry. That's all the Romani I know. Now, let's get down to business."

Johnny put down his fork, a tater tot still impaled on it, and stared at her.

"The first lesson of life is that you will never fit in with THEM," Annie continued, tossing her head vaguely in the direction of the rest of humanity. "Even if you wear the right clothes. Even if you learn the right dance moves. Even if you play the right sports. The best you can expect is that they see you do something cool and you get admired for a few minutes. But you'll never, ever belong. You'll just be like a freak in a side show."

"Funny you should say that," he said.

"It's not funny. It's the hard, bitter, contemptible, unjust Truth." She had to wipe a bit of spit off the corner of her mouth before she could continue. "But you could join a coterie of people who don't care if they fit in. We just care about each other and do cool things and ignore the normal kids. Are you in?" She held out her hand.

Johnny didn't even have to think. If he was going to be totally and absolutely alone for eternity, he'd rather do it with other people. Especially her. "Is there a secret handshake?"

"No." She thought for a second. "But we could do this." She held her hand out and they linked pinkies. "Now go act normal and wait for further orders." She grabbed her tray, stood up and disappeared into the crowd.

A few minutes later Johnny finished the last of his lunch and headed for his next class. He was still thinking about Annie and forgot to hunch over and keep his eyes down. As a result, he found himself looking into a face that was actually smiling up at him. For just an instant his dark eyes met the twinkling dark eyes of a very pretty, nicely dressed girl with silky black hair. Then she was gone. This random meeting did not escape Annie's attention. Naomi Feldman was not yet, but would soon be, the next member of the AC.

At first, Annie had considered Naomi to be the least likely person in the world to join the group. Rather than being a misfit, Naomi seemed just about perfect at fitting in. She was beautiful, classy, smart and mature. It was the maturity, Annie later realized, that was Naomi's undoing. If you crammed a 35 year old fashion executive into a 16 year old's body and made her go to high school, the result would be amazingly like Naomi. The normal kids liked Naomi but Naomi didn't find anything about the normal kids interesting in the slightest. She called them the humdrumniks.

---

[9] "Good day to you." The response is "thank you."

In the afternoon on the same day that she had recruited Johnny, Annie approached Naomi.

"Whaddya think of Johnny?" she asked. "The tall kid in black."

"The black hoodie guy, right? I heard what happened to him in history. Can you imagine? I just felt so terrible for him." Naomi knew what it felt like from personal experience. The last time a guy had made cracks about Jews around her, he'd gotten a face full of very expensive hair spray and an explicit warning about what would happen to him the next time. "He's sorta cute in a gloomy kind of way, don't you think?"

"I guess he's a little cute. I didn't really notice," Annie said, though of course she had. "Let me ask you this. What do Johnny, me, Andy Kanayurak, and Justin Larson have in common?"

Naomi thought for a minute. On the surface there was almost nothing they had in common. Then she got it. "You are all interesting!" Her face lit up. Thank God there are at least a few non-humdrumniks here or I'd be going completely crazy."

Annie was grinning despite her attempt to be cool. She admired almost everything about Naomi and being praised by her was a real thrill. "Well, we'd like to adopt you." In response to Naomi's puzzled silence, she explained, "We've got this coterie – that's like a club, only it's French – and it's only for people who don't fit in, only we do fit in with each other and we'd really like it if you wanted to join but you don't have to if it's not cool with you or whatever …"

Naomi jumped in to save Annie from further embarrassment. "I am SO in!" Then she gave Annie a big, warm hug that smelled of terribly expensive perfume. It made Annie blush.

The final member of Annie's Coterie was, to be honest, a little unusual even by AC standards. He was the only person who had ever gotten a higher score on a math test than Annie. In fact he was the only person who had ever gotten perfect scores on all his math tests. In half the time that it took most of the kids to fail miserably. He would have made a great study partner, except for the fact that he didn't talk.

To be accurate, he did speak occasionally but it was usually one word. And that word was usually messed up in some way, messed up meaning that it wasn't the word that anyone else would have used in the situation but it was often amazingly insightful or at least hilarious. His name was Aaron and he was autistic. Or so they said.

Annie had watched Aaron carefully. On the outside, there wasn't much to see. He looked like a chubby overgrown kid. His sandy hair was too long to be called short but too short to be considered long. It seemed to hang onto his head for dear life, as though it were about to be blown off any moment. When Aaron walked, it appeared as though his mind were somewhere distant and his body was trying to do its best without him.

It was clear to Annie that Aaron's mind was somewhere else. And that somewhere else was really important. She wasn't sure what it was that he was thinking about, but she was confident that he was working on things that the rest of the world was ignoring and that some day the world would be extremely grateful to Aaron.

She'd once made the mistake of trying to find out what was going on in Aaron's mind. She had approached him on the front steps of the school and said, "A penny for your thoughts." He had held out his hand. She had reached into her pocket and pulled out a nickel. He had taken it and handed her back four pennies. Then he turned and started walking away.

"Wait! I wanted to know what you were thinking about."

He had stopped, a puzzled look on his face. "Shrinking about!" he had finally exclaimed. Then he sat down on the steps, pulled out a notebook and started writing. Moments later, he ripped the page out, handed it to Annie and walked away.

Annie had studied that page for a full hour. It was filled with drawings, sketches, numbers, symbols. There was almost nothing there that she could recognize, not even standard math. And yet she had the feeling that she was holding the plan for world peace, an anti-gravity drive, or the reanimation of dead flesh. From that moment on, to Annie Aaron was that piece of paper – a plain-looking exterior with an unfathomable and yet earthshaking meaning. Later on, when Annie had asked him to join the club, he just smiled and said, "Goatery", a word that later became famous among the humdrumniks.

Aaron, it should be said, had never pronounced his own name properly. No one knew whether he didn't hear it correctly, couldn't repeat it, or just had his own reasons. His parents and teachers called him Aaron. Aaron, however, pronounced his name Herring, and pretty much everyone else who bothered to speak with him did as well. He knew darn well what it meant, too, because he insisted on having some every Friday night at dinner.

Herring had completed the Coterie and since he had joined, the group had begun to evolve in ways that none of them had expected. If, as Dr. Tripledoor seemed to think, Annie was in danger, at least she wouldn't be alone.

www.ingramcontent.com/pod-product-compliance
Lightning Source LLC
Chambersburg PA
CBHW081107080526
44587CB00021B/3487